Gray's Talent

A novel by

Roy F. Chandler

Gray's Talent by Roy F. Chandler

Printed in the United States of America

This is a work of fiction. The characters in this book and the situations depicted are the author's creations. They do not exist and did not happen.

Chandler, Roy F.
Gray's Talent

ISBN 1-885633-05-X

Foreword

Nearly all reasonable and civilized persons agree that "War is hell." And it is hell. It is often a worse hell for the survivors than it was for the dead because war can destroy ideas and traditions and beliefs as well as lives and homes and families. War is destruction, and destruction is, generally, a dislocating experience. Why do we do it?

Many philosophers, pondering the question, "Why?" have come to the conclusion that, in spite of being "hell," war is also "Mankind's greatest pastime." History and tradition support this idea, and, sadly, the future will prove it.

This "hell" and this "pastime" can come together in curious and terrible ways. As an example, a veteran of the Vietnam Conflict complained to a TV interviewer of horrible, sweating dreams and flashbacks of shooting a Viet Cong in the face with his .45 pistol as the enemy soldier clung to the veteran's pantleg and begged for mercy. The interviewer asked, "And when did this occur?" The veteran answered, "During my third tour." *Three tours?* Nobody had to go for three tours. He was there because he thought it exciting, and now he realized the hell of it.

Korea was one of the last major conflicts from which the combatants returned home, picked up their lives as best they could, buried the horror, and built no monuments to perpetuate the hell of their memories.

Chandler develops one of his best novels yet around a Korean War veteran who picks up his life after the war, but, later, uses a talent that he discovered in hell to combat the black "pastimes" that can lurk in rotten hearts.

Arthur B. Troup, Jr

Gray's Talent

1

He liked coming upon the house from the field. He first saw an ill-defined mottling of brown and reddish roof tiles that blended within the natural tree colors. When the walls appeared, their white stucco stirred his senses. A deep and shadowed porch ran the length of the building, and he knew that a matching porch protected the opposite side.

Such porches were not common above the Yankee river's broad expanse. The plan was more southern Maryland, where sun beat strongly, and wearied farmers could rest in deep shade with their harried and overheated wives bringing outside any work movable.

In Pennsylvania, homes followed centuries-old Georgian stylings or more lately builder's schemes of toney split levels and stretchy looking ranchers. The few who encountered the tiled and stuccoed home paused for a second look.

Although larger than most, it was not a huge building. Its owner called it a bungalow, although no one else agreed. Three thousand square feet of living space moved the home from a bungalow into at least a small "Wow" class building. An estimator's quick appraisal might figure that at $60 per square foot plus a few grand extra for the fancy roof and the huge porches might bring the house's worth to about $200,000--plus land, of course. The house had actually cost more than one million dollars.

Robert Grayson liked coming onto his house from the field because only from that side could the entire home be seen. On the other sides a photographer could not back off enough to catch it all. The building was tucked into a woods with oaks, maples, and dogwoods left growing almost against the broad roof eaves.

The field view saw the house from its back, and the long rear porch was heavily screened. The front of the home overlooked the river, which could be seen in winter when leaves were gone.

looking into woods, the front porch boasted a half dozen well-used country rocking chairs scattered along its fifty foot length. The porch columns supporting the twenty-five or so tons of roofing tile were square and tapered from top to bottom. Six feet apart, they looked clean and strong, as if they could support anything piled on. They also allowed porch sitters to feel a part of the forest that encroached with an occasional leafy branch entering the porch itself.

The house was nearly new, but without the empty rawness of an undeveloped yard it appeared to have been in place for decades. Comfortable, might describe the home. A man, a couple, or a family could like living there, but at a million dollars?

Robert Grayson could smile at the expense. The house money and enough more had come easily; he had stolen it.

+++

Bucky Larson sat motionless, his True Tree camouflage blending well with the forest. The scoped rifle had been daubed with brown and green so that only the telescopic sight's objective lens remained clear. Larson sheltered its possible reflection with a taped on cardboard sunshade made from a toilet paper roll that extended four inches beyond the scope's lens.

Only Larson's eyes moved behind his net-like Sniper's veil. Backed against a tree with laurel branches above and around he was hard to see. He had better be. His quarry was one of the best and could not be taken lightly.

Larson saw the man coming, a still distant figure barely caught through intervening forest. Larson swore silently, his finger remaining poised above the rifle's safety. The walker became lost behind trees, but he would probably be following the dim path that wove through the woods. He would appear again only yards across the small clearing Larson faced. Maybe . . . Larson sharpened his senses. His mind should not wander. If he got the shot it would be quick and only momentarily open. Others had tried and failed. Some doubted it could be done, but Larson was doing it right. He had planned long. He had come in before daylight, and there had been no carelessness, nothing left unconsidered. Luck always played a hand, and Larson swore again, eyes searching as he listened for the man's approach.

Something hard pressed the side of Bucky Larson's neck, and he jerked reflexively.

"Bang!" The voice held amusement, and Larson tried to accept that he had been crept up on without even a hint of it happening.

Bucky flipped his face veil aside and tried to grin as the man he had seen walking stepped from behind him. He knew his voice held astonishment, but it had to. He could not believe it.

"How in hell . . . ? I saw you way over there, Mister Grayson, just walking along. Then all of a sudden you're right behind me sticking something in my neck. How'd you see me? How did you get from there to here? I mean, I was watching hard and . . ."

Grayson extended a lean-fingered hand to help the hunter to his feet. He hauled, and Larson came erect, leaves and forest humus falling from his camos, shaking his knees to loosen joints and muscles stiffened from long sitting.

Ignoring Larson's question, Grayson asked his own. "I guess you were after Old Bart?"

At Larson's nod, Grayson chuckled. "That has to be the oldest and smartest squirrel in the woods. Men have been after Bart since I got back here."

"You ever see him, Mister Grayson?"

"Well, Old Bart is supposed to be very small and coal black. I've seen a squirrel like that. Of course it might not be the same squirrel each time. He was always way out and moving fast."

"That fits Bart's description. A friend of mine, Mike Thompson, hunts him like I do. Mike says that all you ever get is a moving glimpse."

Grayson's smile was internal. This was his land, and he knew every twig. He knew Mike Thompson's favorite hide, which was a lot closer to Bart's territory than Bucky Larson's spot. Hell, he knew where Old Bart lived, but it had taken a lot of careful stalking to find Bart's tree. Larson would have to find the black squirrel on his own.

Bucky came back to his question. "How did you work in behind me like that, Mister Grayson? I still can't believe you could do that."

Grayson could have said, "Because I have spent my life learning how," but he chose to point out his route.

"I just ducked into a little swale that runs back there and came up walking quiet."

"Yeah, but how did you see me at all?"

"Bucky, I walk through here all the time. I happen to know that this tree does not have a solid blob at its base. I figured it was you hunting, and seeing I had already put any squirrels into cover I thought I would just give you a start."

The explanation was close enough. He had seen the too solid "brush" at the tree base, but until he had gotten close he had not guessed the hunter's identity. When he had touched Bucky Larson's neck, his other hand was still returning the pistol to its holster at the small of his back.

Robert Grayson could not take camouflaged riflemen as lightly as many might.

Grayson moved the conversation ahead. "That's an old Sako you are using, Bucky. What's its caliber?"

"It's a .222 Remington Magnum, Mister Grayson. It's an old cartridge, and you don't see it around much."

Larson offered the rifle for examination. "Careful, there's a round in the chamber." Then added, "The scope is even older than the gun. It's a Weaver K-6."

Grayson took the rifle with easy familiarity. "Well, the .222 is old, but it is plenty for this hunting. It would be a miracle to get more than a one hundred yard shot in this woods, so your scope is just right, too."

Shouldering and aiming the rifle, Grayson thought a moment, "I used to load the .222. I was out west then and used it to hunt prairie dogs. Until the .222 came along I had a Hornet and a .218 Bee. Ever use either of those?"

Bucky liked the gun talk. He barely knew Grayson, but he had used his official car and had been in uniform when he had asked for permission to hunt on Grayson's land. That let Grayson know that he was a highway patrolman and a responsible person.

Grayson had said, "Sure, any time you want. Another officer already hunts here, and I'm out in the woods a lot, so don't be surprised if I ruin your hunting now and then." Bucky had not seen Grayson again, until now.

Grayson was pretty old, in his sixties Bucky suspected, but he was lean and fit looking, and the way Grayson had slipped up from behind was amazing. Bucky gave Grayson closer attention:

--about five feet ten inches tall, maybe one hundred and seventy pounds, graying hair, blue eyes, no glasses, moved light on his feet, dressed in khaki pants, tucked in polo shirt, light sweat shirt, carried a holstered handgun under the sweatshirt.

A Pistol? The law officer's eyes squinted. Carrying concealed on your own property was legal for anyone but highly unusual. Of course, off-duty cops and retired law officers--especially those who had served in mean-assed big cities almost always carried a gun. Bucky felt undressed and a bit insecure without his own off-duty pistol.

Maybe Grayson was a retired police officer, but they usually made the fact known to other cops, and Grayson had made no mention of what he had done for a living.

A law enforcement officer could be made uncomfortable by a friend or an acquaintance flaunting a gun-carrying regulation. What did you do, ignore it, warn, arrest your buddy? Remind might be a good idea.

Bucky took his rifle back, unloaded the chambered round, and, gripping the empty rifle between his knees, he shed his too warm head gear. He made his voice casual.

"You a retired Pennsylvania cop, Mister Grayson?"

When he answered, Grayson was facing away, looking through the woods and across the field to where the ridge of his house roof could be seen. His answer was unconcerned and in two parts.

"Retired federal, Bucky. I've got a permanent permit to carry concealed."

Grayson did not offer his permit, and Larson was not about to ask to see it. Bucky felt his face heat at the way Grayson had read his intent. He wondered how Grayson had understood that his question concerned the pistol concealed beneath the sweatshirt. He would like to get a look at a federal permit; he had never seen one. Permanent? He wondered what Grayson's job had been to qualify for that kind of document.

Bucky chose another approach.

"I saw your pistol range. Don't you shoot rifles anymore?"

For the moment, Grayson ignored the patrolman's question and asked his own.

"You calling it quits, Bucky? The sun is getting high and squirrels won't be moving."

"Yep, my stomach's calling lunch, anyway."

Grayson led the way through the woods toward the dirt road where Larson's vehicle would be parked. Then he handled Larson's question.

"Anybody who can shoot a pistol will do all right with a rifle, and I'm too old to hike back and forth to check rifle targets."

Bucky said, "I didn't see any targets and only a few empty cases, but your backstop is plenty shot up."

"I bring in my targets. I don't want anyone seeing how badly I shoot, and I reload most of my cases. Saves money, and I still enjoy working up loads."

Larson said, "I looked over the leftover cases. Someone's been shooting .45s, .380s, and 9mm. A lot of .22 Long Rifle, too."

"I doubt anyone else has been shooting there. I leave some bad cases, all of the rimfires, and I miss others. You know how it is, empty an eight shot magazine and you will be lucky to recover half a dozen empties. Some get trampled, others just disappear. You know how it goes."

"What are you carrying now, Mister Grayson?"

"A little flat sided .380. What have you got?"

Bucky was a little embarrassed. "Seeing I have my rifle I only brought my .22 wheel gun."

"Must be a North American Arms .22 Magnum. Everybody has them these days."

"That's what I've got. Not a good choice for out of doors now that I think about it."

They reached Larson's truck, a top of the line Ford F-150 with plush seats and automatic windows.

Bucky opened the cap and dropped the tailgate. He wiped the Sako down and secured it in a foam lined Pelican case. Grayson watched silently, but Bucky felt the man judging how he handled his weapon.

Bucky closed the rear end and opened the driver's door. He smiled, "Well, Old Bart makes it through another day."

He climbed behind the wheel, started the deep throated engine, and they both listened to its powerful rumble.

Larson opened his window and closed his door. Fastening his seat belt he said, "I'll buy you a lunch at Steve's if you're interested, Mister Grayson."

The older man's answer was friendly. "Not today, Bucky, but I'll take a rain check. I ride my Harley out that way pretty often. I'll look for your truck."

Larson drove away thinking about it. A Harley-Davidson! Grayson looked more like a sports car type, maybe an MG, or more likely an XK120 Jaguar.

An interesting man, that was for sure. Carried a piece and moved like an Indian. How in Holy hell had Grayson slid in behind him that way? Robert Grayson. Larson guessed he would ask around about him.

Cutting back through his woods, Grayson thought only a little about the young police officer. A sharp young man with a trained eye. That was clear. Grayson's .380 lay tight in his back; you had to be looking to detect it.

Grayson altered his course and came onto his pistol range. The shooting spot was primitive, simply a one hundred yard long path cleared to a small embankment that would absorb bullets. he removed a block of three inch Post-it notes from a sweat shirt pocket and stuck six in a row along a shotup target-holding board fifty feet down range.

Grayson walked away five steps then snapped around drawing his .380 as he turned. His body crouched, but his arms straightened to eye level. The pistol cracked three times, shifted to the empty hand and fired three more shots.

Grayson popped the magazine with one live round remaining in the pistol's chamber. A full magazine came from the sweatshirt pocket, and the pistol was again fully loaded.

He gathered his spent brass and his Post-it notes. He examined them as he walked. Four were solidly hit, one had a bullet's half moon cut along an edge. One had been missed. Not missed by much, Grayson expected.

2

The house had started as doodling on a scratch pad. The stranger in the adjoining cubicle had identified himself as Dick who had a degree in architecture. With little to occupy them, Grayson's house plan became an attractive time filler.

The simple sketch became careful drawings and eventually developed into a stack of complicated blueprints. Ideas beget concepts and many were included in the evolving structure.

Grayson and Dick did not discuss their occupations, but Dick understood Grayson's requirements more than most could have. Their work, each expected, at least paralleled the other's.

Dick said, "Geothermal heat pumps will give you the best heat and air conditioning. You can run your heat exchangers horizontally if you have a big field, or you can put them down wells if you are short of land." Grayson had a field. He brought plats and aerial photographs and they spent time positioning the house exactly right.

The architect called Grayson, "Gray." Nearly everyone did.

"Gray, you'll want three tunnels. One to the river, and one out each end, but the fourth side, the field side, presents an expensive problem. It is one hundred and twenty yards to the woods edge, and you would want your entrance well inside. Too far to tunnel, I think." Gray solved the problem another way, but that was after Dick's desk had been cleaned out, and a man called Phil sat there. Dick never resurfaced.

Fortunately, the house plan was virtually complete, and Gray began building. Sometimes he wondered about Dick. The man could have retired or been transferred. Dick might even be building his own home based on their plans, but Grayson doubted it.

A local man excavated Gray's cellar and concreted the footers and floor in one monolithic pour. He had groused over the silliness of three foot wide footers and all of the rebar in both the footers and the ridiculously thick six inch cellar floor.

As if that were not twice what was needed, additional reinforced footers crossed under the floor where interior walls would later rise. All of the rebar was 5/8 inch and epoxy coated. It was laced in and crisscrossed like a bank vault, and where there would be walls, the reinforcing bars projected upward more than a foot on sixteen inch centers.

A big company with a lot of cement trucks formed up and poured the cellar walls, the main floor, and all interior walls. The house walls were capped at the square with a twelve inch thick flat roof. The owner insisted on one continuous pour--no cold joints would be tolerated. The pour was a challenge because of the crisscrossed rebar and plumbing and electrical conduits running through everything and the complication of large window openings.

The walls were twelve inches thick, and they were pouring a special 3500 psi concrete with fiberglass added. The workmen decided they were building some sort of a government bomb shelter, but they could not explain the windows. Holes in a bomb shelter did not make sense.

Twenty laborers were on the scene the day of the pour. There was a lot of cursing and ordering, and the owner hung around looking closely, but not interfering. When the crew finished there was satisfaction in a tough job done right. Men cheered, and the owner (nobody seemed to know him) popped for a few cases of iced beer.

Two months later, after the cement had set up well, a second story was added. It was smaller than the first but just as heavily constructed.

The contractor and Mister Robert Grayson, who provided the plans and paid the bills, were pleased with their arrangements. The work could be done right because Grayson was willing to pay a fair price. Best of all, Grayson had asked if he might pay in cash. Straight faced, the contractor had agreed. Cash was like a twenty percent bonus. A big portion of the money would never hit the books and would therefore be tax free.

When Mister Grayson noted that he would not need receipts the contractor vowed to do the best job possible. Grayson obviously understood the real world. Receipts could come back to haunt a tax skimmer. Grayson was the kind of customer a company dreamed of acquiring.

The bomb shelter's walls had been raised on the inside edge of the foundation's massive footers. The contractor had wondered at the specification and was curious about a second row of upright rebars protruding from the footers outside of his walls.

Only weeks after the second pour had been finished Grayson was back with more plan pages. The construction requested was simple enough. Do it all again!

Foam insulation ten inches thick had been glued to the outside of the concrete bunker. Now, a second bunker, with the same reinforced twelve inch thick walls was to be built tight to the insulation.

Incredible! Walls and roofs would be nearly three feet thick. What a cash contract!

The work went smoothly with many smiles, and at completion there was the now ritual cold beer celebration.

Some over-large stainless steel and nylon rollers between the walls at the window openings explained how the bomb shelter could be closed up. The workers figured some heavy steel plate would move back and forth on the rollers sealing the windows against flying bomb parts. One hell of a structure, the workmen agreed.

Robert Grayson brought Ronnie Barrett to see the concrete fortress. Barrett built wooden houses, and he was a builder with enough imagination to handle an unorthodox job.

Gray unrolled his plans. "What I want you to do, Ronnie, is to build a house right over the cement one. The roof slope will hide the square upper walls, and the place will look like a common home with attic rooms. The tile roof will add a lot of weight so make your work strong. Do not skimp anywhere. See these posts supporting the roof?" Grayson's finger pointed at the plans. "You can frame them out in wood, but inside they will be five inch solid steel shafts. Not pipe, Ron, solid steel. That is the kind of construction I want. Can you do it?"

"Oh, I can do it, Mister Grayson, but I've got to ask, Can you afford it? This is going to cost big bucks. Too much for me to even guess at."

"I can pay, Ronnie. It'll be cash, on time, whenever you call for it."

The job took awhile. Inside, all concrete walls were either dry walled or paneled.

Thousands of feet of black plastic pipe were buried eight feet deep in the field. Filled with antifreeze they fed a pair of heat pumps that would work no matter how cold it got.

The heat and air conditioning duct installers found their openings in the concrete walls exactly where they were supposed to be, and the electricians discovered three inch conduit providing easy passage for their wires.

A team came from the city with the steel plate that would slide on the between wall rollers. Barrett's carpenters stopped work to watch the installation and to lend a hand where practical.

Steel plate did not adequately describe the sliding shutters. Each plate was three quarters of an inch thick, and if tapped, the metal rang like a gong. Armor plate, the installers explained.

Almost impossibly heavy to maneuver, the steel was edged in, dropped, turned, and lifted, but once in place, the weight rolled easily on the nylon rollers. The house could be buttoned up with little effort.

The builders had to wait for the house windows. They too came in from special construction. Until he saw the windows, Ronnie Barrett had believed the freight company foolish to demand that a forklift be on hand to unload. It would be, after all, one hell of a window that four men could not muscle around.

Barrett had never seen pre-built steel-framed, double-paned bay windows, two of which were over ten feet long. At least he had not seen such windows fitted with two layers of five eighths of an inch thick bullet proof glass. Without a forklift, nothing could have been budged.

Planning and construction blended well. The windows slid in place, and their one quarter inch thick steel flanges four inches wide and pre-drilled in various patterns allowed the frames to be securely bolted to the walls. Steel, right-angled triangles came with the shipment. They were lag screwed to the walls beneath the windows, lest the tremendous weight eventually sag the wooden studding.

Ronnie Barrett decided to double up on all of his wood framing around the bay windows and to add his own angled two by sixes under the bay windows. He could then fill that space with insulation and close it up, hiding the massive window supports.

Concrete roof tiles came up from Florida. Each weighed nine pounds, and Barrettt used five thousand, five hundred of them. The roof loading was huge, but the construction plan allowed for it.

Grayson came often to see his home go up. He did not rush the construction, and building such a one of a kind home ate up years.

The tunnels were money but not time devourers. Because Grayson would not allow large trees to be destroyed, the tunnels curved and dog legged. They ran from the house basement in three directions. Each was made of six foot long sections of rectangular concrete pipe manufactured for big city underground piping and wiring. The tunnel sections were six and one half feet tall and three feet wide. They locked together with nylon bushings that sealed completely.

Grayson used the manufacturing company's installers, and they made light work of the daunting project. Accustomed to working amid urban obstructions, the workers only ritually groused about trees being in the way. Within a week the tunnels were covered, the ground thoroughly compacted, and the normal forest floor mix of needles, leaves, and rotting material restored. Fast growing trees were planted on top of the tunnels, and only dark entrances betrayed their presence.

A steel door closed each outside entrance, and a second door within each tunnel further sealed the passages. At the cellar end, vault type doors hermetically sealed each tunnel.

Less obvious than the steel doors were electric wires and a copper pipe that ran on opposite sides of each tunnel's upper corner. The wires powered lights and unobtrusive motion and sound sensors. The pipe had regularly spaced shower-like heads. The plumbers who installed the piping understood the system to be a fire fighting sprinkler that would be hooked up later on. It did not occur to them that a disabling or toxic gas could just as easily be sprayed through their pipes into each tunnel.

Grayson had begun building his home three years before he retired, Early retirement from some occupations were common and often necessary. Infantry soldiers usually packed it in after twenty years or shifted to less physical duties. Police officers quit young or took desk jobs. Fire fighters left just as early.

If they survived, men or women in Grayson's line were usually put out to pasture a decade earlier than Grayson's age. Unfortunately, individuals with grudges often sought out and prematurely terminated those deserved retirements.

Robert Grayson had been active far longer than most, but, of course, he was far better at what he did. Gray had a real talent, and although they did not know who he was, most of those who used his abilities disliked losing such an exceptional craftsman. But time moved on, and the demand for Gray's services diminished.

Grayson recognized the approaching gold watch, and leaving the business had been in his mind even when he had first begun his house doodles.

To go home, home to the small town plastered against the side of the Susquehanna, to turn off Main Street that paralleled the river and go uphill past the garage on the corner of High Street, and to keep going almost to the ridge top where the road ended in a Tee, that was the way. Then turn right and travel back downhill along a curvy road to its end. There Gray's land lay, just above the river with a field alongside. He had a hundred or so acres, but he was surrounded by woodland, and there were no houses along his road or within half a mile in any direction.

There were other ways into Gray's land, but no paved ones. If an enemy with a too long memory intended to ambush him in the open, the road in seemed the most likely choice. Gray could look out for that. More probably, a vengeful individual would attempt to nail his hide while he was at home, possibly while he slept. Gray's house was assurance that such an attempt would be unsuccessful.

And the field, across which a Sniper could shoot with expectation of hitting within an inch of his chosen spot?

First, Gray had prepared a perfect spot for such an attack. The ground was elevated a hint so that field grass did not obscure the house windows. Foliage disguised the hide, and there were natural looking bramble free routes of approach and withdrawal.

As soon as his cellar was completed Gray rented a Ditch Witch and dug a five inch wide trench three feet deep from the house to the hide. He ran a camouflaged wire up the far side of a tree and placed a motion detector almost above the Sniping position. He planted his explosives in three packs eighteen inches deep under where a shooter would lie. Two of the charges would make certain that either a left or right hand shooter (or a Sniper with an observer lying by his side) would receive the full effects. The last charge was seven feet behind the shooting site, where a third man would most likely position himself.

Grayson buried his wire and graded the Sniping spot so that it had to be irresistible. He had his field plowed, harrowed, disked, and replanted in grass.

If his detector sounded, Gray could enter his attic, raise a certain roof tile, and look through his 10 power telescopic sight directly into the Sniping hide.

Gray's counter-Sniping rifle was a custom Remington, bolt actioned .44 Magnum. The .44 handloaded cartridge moved only slightly above the speed of sound, and Gray's eighteen inch long Maxim type silencer reduced the rifle's report to a barely detectable plop. At only a little more than one hundred yards the effect of the heavy and large diameter .44 bullet would be terminal. The rifle lay ready The sandbag rests were also in place. Grayson had only to chamber a round and make his shot.

Occasionally, a passing animal touched off the cross-field motion detector, but that was a rarity. Each time, Gray looked, made certain, then reset his alarm.

A pair of wires separated and mounted to a roof rafter were each capped with a plastic protector. A US Army hand operated explosive generator with alligator clips was placed beside the silenced rifle. If more than one adversary lay in waiting, Gray could use his buried explosives. He had carefully judged his charges. Death would be certain, but he would not have to gather up pieces.

Robert Grayson could prepare all of these things because he had a great deal of free time. He had stolen more money than he could sensibly spend, and he had had access to all of the nifty hardware required. Of course, Gray also possessed the necessary skills and knowledge.

The bottomless armament pit closed to him when he left the company, but, as with his house, Gray had already prepared. If a group chose to wage war against Robert Grayson, they had better arrive with heavy weaponry.

While he was active, Gray was to some degree protected. To attack him was to attack the agency. Most opposing groups around the world might consider that to be a poor risk. Of course, no enemies were supposed to know about Robert Grayson's activities --but who could really be sure?

Once retired? The security balance changed. The Graysons of the world were numerically very few, and exactly who they were or even what they might look like was usually unknown. But supervisors changed and names or details could be bandied about. Although no Grayson types were completely abandoned to their fates upon retirement, the personal danger factor often rose dramatically. Robert Grayson had long decided that he would not be listed among the suddenly deceased.

For two years following his official retirement, Grayson traveled about, ending the possibility that old adversaries might come to square ancient insults.

3

Gray's motorcycle was a customized Harley-Davidson softail. Only the black Corbin seat and the blue painted tanks and rear fender interrupted the flow of show quality chromed metal extending from wheel to wheel. In sunlight the bike glittered like expensive jewelry, and that was fitting because a sharp Harley was worth attention-grabbing money.

It was often repeated that a Harley-Davidson, any Harley-Davidson, was the only form of motorized transportation that you could purchase retail, use for years, and sell for a lot more than you paid for it. For half a decade there had not been enough Harleys around. Harley motorcycles refused to die, and junked out basket cases were routinely restored to show class condition. Although the Milwaukee company increased production to four hundred bikes a day, its York, Pennsylvania plant remained sold out for a year ahead.

Harley-Davidson owners carefully locked their cycles because stealing Harleys had become as common as heisting cars. A stolen motorcycle was usually disassembled in a garage and the frame replaced. Any distinctive custom touches that might identify were removed and spread among other bikes. The paint job was done over, and within a day or two of the theft, at the expense of a few hundred dollars, a "reconstructed" perfectly registerable Harley worth ten thousand or more dollars emerged.

There was concerned argument among riders over how long the Harley fad would continue. Kawasaki, Honda, and Suzuki were offering Harley clones at lower prices that were motorcycles of far more modern design and that to an inexperienced eye looked like Harleys.

Despite the Japanese efforts Harley popularity held. Americans and thousands of Europeans enjoyed love affairs with their Harley-Davidsons. People who did not even ride a motorcycle wore Harley clothes and had "Live to Ride--Ride to Live" decals on their car windows. Men and some women had Harley logos tattooed on their skin. No one displayed similar Coca Cola or Marlboro body decorations. Harley-Davidson (always with a dash) was an almost inexplicable phenomenon.

Robert Grayson was not that much into the Harley-Davidson game, but he had learned on a Harley, and he had stuck with the company when its undeserved outlaw biker image had nearly produced company meltdown.

Even that part of Harley-Davidson history was astonishing. A Marlin Brando movie about bad-assed bikers in black leather became a national hit. The fact that Brando rode an English Triumph motorcycle in the movie made no difference; Harley took the hit. The Hells Angels Motorcycle Club rode Harleys, and everybody thought they knew how awful those people were. A fistful of "B" rated biker films followed Brando's success, and Harley-Davidson had an image hole to try to climb out of. A double handful of the faithful hung on, continuing to ride their Harleys, and Grayson was pleased to have been one of them.

Grayson's ride was a 1988 model, but year meant little to a bike's value. After market modifications were what counted. Gray's machine had been White Brothers lowered, and KuryAkyn and Arlen Ness pieces were bolted on. Those names rang strangely to non-riders, but they were powerful talismans within Harley circles.

Gray's engine had a Screaming Eagle cam and an S&S carburetor, small modifications that boosted performance a touch. The twin Python II exhausts created just the right Harley, V-Twin thunder and the deep thud of *potato, potato, potato* at idle.

Grayson rode his motorcycle in good weather and when he was not in a hurry. He avoided being on the bike with deadlines imminent. Hurried riders got into trouble, and riding on wet roads and in rain accomplished only two things. They guaranteed discomfort and hours of work again making the Harley shine like it should.

Coming up-river Gray saw two patrol cars pulled in alongside Steve's restaurant. Troopers usually parked in plain sight lest John Public suspect they were goofing off. Each police car had a number along its side. Number seventeen was Bucky Larson's.

It had been a week since Gray had found Larson hunting Old Bart, and it was lunch time. Grayson decided to visit. Grayson signaled and began easing across halted traffic. A driver waved him through, and he got into the right turn lane. A motorcycle had some advantages.

Gray pulled up in front, alongside a powerhouse of a pickup truck with huge wheels and a six spotlight bar above the cab. He locked his ignition and stepped into the restaurant.

Grayson's eyes swept the place with disguised intensity, automatically logging all that he saw. Customers were few. The pickup's occupants were probably the three at the bar. The small skinnier one was just easing onto his stool. Faded out tattoos, a pigtail, and heavy boots hinted that he might be interested in motorcycles. Gray expected the man had gone to the window to see what machine had made the good two cylinder engine sounds.

The troopers had a window booth. Bucky raised a hand, and Gray headed for them.

Gray only half caught Larson's introduction of his companion, shook hands, and chose to sit where he could see most of the restaurant.

Bucky said, "Now that's a motorcycle!"

Grayson agreed, ordering coffee and a tuna salad on rye. "I've got fifty thousand or so miles on it, but the engine runs strong."

"Who does your work?"

"About all it needs is periodic maintenance. I run down to Jim Howell, outside Carlisle.

"They've got a good reputation."

"They do well by me, and I like Jim."

"You know about the Blue Knights, Mister Grayson?"

"Blue Knights?" Gray's attention was drifting a little.

"It's a police-only motorcycle club. It's all over the world."

"Do you fellows belong?"

Bucky answered, "Not right now. Bill doesn't ride, do you, Bill?" A head shake. "I'm holding out until I can afford a Harley."

"That'll be awhile, Larson." Bill, last name unknown, sounded certain. "My brother put out over twelve grand for his."

"I could start with an old iron head Sportster, then trade up when I get the money."

Gray liked the motorcycle talk. He also liked cops. Unlike some bike enthusiasts, cops tended to be clean, sober, and headed in good directions.

Only by a long stretch of imagination could Grayson count himself as a police officer. A federal permit to carry did not necessarily imply law enforcement.

Bucky said, "We've got a Blue Knight chapter in Harrisburg, Mister Grayson. You ought to join. It'd give you someone to ride with."

Grayson's mind was partly on the trio at the bar. Their heads were a little too close together and their voices too guarded. Gray had looked for such signs for too many years not to notice. He answered Bucky's suggestion.

"Maybe I'll look into it, but it might be better to wait until you find a bike. Then we could join together."

Bill snorted. "Bucky won't ever get around to it. His idea of excitement is to lay on a rifle range trying to make all his bullets go into one hole."

That interested Grayson. "You practice a lot, Bucky?"

"Quite a lot, but Bill's talking about my Sniper gun, not that .222 you saw me with."

"Bigger caliber, I suppose."

".308 Winchester. That's pretty standard for police Snipers."

Gray said, ".308 is a good choice. Your Sniping is mostly under one hundred yards. I suppose you use Federal 168 grain match ammo."

"Yep, everybody does. You interested in Sniping, Mister Grayson?"

"Not anymore."

"You still have a rifle?"

"Yes, I have a few. Come around some time and I'll show you. Two in particular might interest you."

"What do you have, Mister Grayson?"

"Well, I've got a .50 caliber bolt gun that's nice."

"Oh man!"

"But the one you would like best is my Iron Brigade rifle."

Larson's eyes popped. You've got a *Chandler Sniper?* Man-o-man, that is THE top of the line. I've only even seen one. Tim Cameron, a police Sniper down in Maryland has one. That rifle shot half minute of angle out to one thousand yards."

"What in hell is a half minute of angle?" Bill was getting behind in the discussion.

Grayson said, "The rifle will put all of its bullets in a five inch circle at one thousand yards."

"Holy hell! You could make head shots at ten football fields distance. God, head hits at over half a mile."

Gray grimaced. "Not likely, Bill. All that means is that the rifle could do it if the shooter, the weather, and a dozen other variables were exactly right."

Bucky said, "I'd trade all my guns for a *Chandler Sniper.* What number is yours, Mister Grayson?"

"Number seventeen." Grayson grinned.

Larson groaned, "I can't believe it, that's my car number. I'll buy it off you, Mister Grayson. Name a price."

"Not today, Bucky." The small man at the bar was leaving, and his companions were getting up. The small man's eyes swept across Grayson's but kept moving. The man sauntered too casually, and Grayson's senses keened.

Bucky was speaking about shooting, but Gray's attention was on the two bigger men. They were big with beer fatted bulk. As they rose, each man's eyes drifted Grayson's way. Unnatural, and Gray did not like it, but unless a new model had been produced, these local hulks were not the killer types Robert Grayson looked out for.

Gray leaned to ask Bucky if he knew the two when the motorcycle engine bellowed. The rumble meant nothing special to the highway patrolmen, but Gray knew his engine's sound.

He came up standing. A glance through the window showed the small man straddling the

Harley. Gray's eye caught the black of a long jump wire skirting the ignition.

Bucky said, "What?" but Gray was already moving.

The two large bodies blocked the door in seeming innocent conversation, but Gray went past them without pause, seeming to almost glide between tables and out of the restaurant's side door even as the Harley pulled away.

Bucky Larson was not that far behind. He had been slower in catching on, but when Grayson moved it had struck home. A motorcycle was being stolen right under their noses. Incredible!

Larson hit the front entrance, fighting his way through the men blocking it. A foot tripped him, and a body shouldered him into a wall. He stumbled through, but the biker was already out of the lot. He caught a glimpse of Grayson sprinting across the next lot. Then Larson was in his car and cranking his engine.

He had time to marvel. Grayson was trying for the traffic light. God, that was fast thinking. If the thief got a slight delay, Grayson might just get him. Larson recalled the .380 he had seen at Grayson's back. He wondered if the older man would take the shot--if he got one at all.

Bucky hoped not. If Grayson shot the thief all sorts of official hell opened. These days it was tough to get away with shooting anyone not threatening your life. Larson hoped Grayson could just hold the bike thief at gun point.

Gray ran hard, an exhausting sprint with only a guesswork chance of success. If the thief made the highway he would go like hell, then zip into a side road and into whatever cellar or old barn the three had planned on. The troopers would be way behind and would never catch a glimpse of the stolen cycle.

Gray angled across the corner lot hidden from the intersection by a mound of imaginative plantings. He could hear his motorcycle on the side street, so he still had a chance. Man, he was too old for this! Gray pushed harder.

He saw the motorcycle caught by traffic with the rider trying to squeeze past waiting cars, edging, feet down, along the right hand curbing.

Gray came in fast from the rider's right rear. A pair of older women pedestrians reeled back as the cycle crowded their curb. The bike got around. The rider looked left to judge oncoming traffic, and Grayson caught up.

A last long stride, a short hop with a leg slung across, and Gray thumped onto the pillion seat behind the rider. The Harley swerved and wobbled from the weight shift, but Gray's right forearm clamped across the thief's throat, and his left fist gripped the rider's skull and lank hair.

Gray's words were clear if a tortured and air struggling gasp. "Pull over or I'll break your damned neck."

His chin forced high by Gray's forearm, the rider still swore vilely and tried to accelerate. He opened the throttle, and his left hand came off the clutch. The Harley surged ahead, and without hesitation Grayson clamped and twisted. It was a practiced move and the thief's head was wrenched impossibly sideward and reversed with a wicked snap.

The pedestrian close beside the motorcycle could not hear the thief die. The engine thunder masked the popping crunch of rupturing vertebrae, but with dead hands on the grips, the Harley wobbled uncertainly, struck the curb, and crashed onto its side.

Gray rolled clear and stumbled to his feet. The thief's leg was trapped beneath the engine, but the dead body would not notice.

The Harley choked itself out as Larson's patrol car bucked across the curbing and skidded to a stop, lights flashing, blocking highway traffic that could have plowed into the accident.

Gray sat down on a ground mound allowing his stressed body to recover. His wind came back, but the ache and tremble in his legs took longer. He had skinned a knee and rolled through something greasy, but nothing felt bent or broken. By the time he got himself straight Larson had the accident scene ordered and secured.

The second patrolman arrived with the thief's accomplices confined in his backseat. A local police car pulled up, and traffic was directed around. In the distance an ambulance wailed, and Bucky talked with his women witnesses and one car driver who had hung around.

Larson came over and squatted by Gray.

"By God, Mister Grayson, I didn't think you had much of a chance." He nodded toward the motionless thief.

"He's dead. Looks like he broke his neck when the bike went over. You sure you're all right?"

"Yeah, I'm fine, but I'll ache tomorrow. A man shouldn't sprint like that after sixty."

Larson was admiring. "You looked like an Olympic athlete crossing that lot." He paused, "I need to hear exactly how it went when you caught up."

"The thief got stuck by cars waiting for the light. He edged up the right side and was just starting out when I jumped on the back. He lost control, and over we went. I rolled clear easily enough. It looks like he didn't."

Bucky asked, "You have hold of him at all, Mister Grayson?"

"Yeah, I was grabbing at him, but the bike went over so quickly I'm not sure what I reached." Gray's eyes did not fall away, and his voice remained clear and certain.

Believable lying was easy. Grayson had done it for decades.

The Harley was a mess, but Grayson appeared little concerned. Jim Howell's Harley-Davidson came with a trailer, and the motorcycle went to the shop.

That Grayson was little disturbed by his cycle's damage was explainable. "It's insured, Bucky. Hell, I could have let the bastard steal it and not lost anything. I didn't run him down for the money."

Bucky could understand that. If a man could do something, he should not let criminals walk on him.

Grayson went on. "I hate dirty little shitpokes that just grab and run relying on bald-assed luck to get away. I swear, their stupid arrogance pisses me off more than their stealing."

Bucky nodded, "I know what you mean. This guy sure wasn't a master criminal. Now he's dead and on his way to the morgue."

If Grayson was unbothered by his bike's damage, he seemed to care even less that the thief had died. The indifference was unusual, and Larson judged that Mister Grayson had seen more than a little of death and dying.

Bucky said, "I'll drive you home. There'll be a written statement, of course."

"Yeah, there's always a written statement, and the insurance company will also bust my butt before they pay up." Gray mused, "I wonder if there has been a time in history that an insurance company just said, 'OK, sir, we'll mail you a check today.'"

"Not that I ever heard of."

Grayson said, "I gather those two ugly lumps that were with the thief never saw him before today."

Bucky laughed, "I can tell you've been in this line of work, Mister Grayson. Yep, Bill said they claimed to have picked the thief up hitchhiking about ten miles back. They said the guy thanked 'em and was going his own way."

"Why am I not surprised? And, of course, those two just happened to be standing in the doorway blocking anyone trying to pursue."

"Pure coincidence, Mister Grayson. We'll never tie them in."

"Who cares, anyway, Bucky? Even if you started shooting those kind of slobs, you would run out of ammo, and they would still be all over town."

Bucky thought about the crime. "What we ought to do is plant locator bugs on a few bikes, set them out to get stolen, and follow them to the chop shops."

They listened to radio chatter as Bucky drove up the river, but nothing important came over.

Bucky said, "I can't get over how fast you reacted, Mister Grayson. What federal agency did you work for?"

"A branch of the Congressional Budget Office, Bucky, I was just a paper man. The job gave me time to play a lot of sports. I liked handball and racquetball in particular. You play those?"

"No, I like the outdoors stuff."

Larson was not sure he had gotten a real answer, but his question had been turned and would be hard to work in again. Maybe he could ask why a paper pusher carried a gun after he retired? Larson mentally shrugged; for all he knew all ex-budget guys were required to pack a piece.

Feds were far removed from state and local law enforcement. The few contacts troopers had with the FBI or BATF were usually frustrating with inter-departmental distrust and self-serving maneuvering. FBI agents seemed to hide their cards and to treat state and locals with patronizing toleration. Troopers too often found federal agents inflated with self-importance while blundering about in unfamiliar geography and conditions.

Larson had limited respect for the Feds. The CIA had no in-country activities, and Bucky had no contact with them. Larson actually knew little more about intelligence work than did other TV watchers and novel readers, but he supposed the CIA, BATF, FBI, and perhaps the budget people were all much alike.

Bucky dropped Grayson at his driveway. The older man groaned a little getting out.

"My plan is to sit in the hot tub, then sack out until I feel human again. I'm too old for this bad guy catching, Bucky."

Larson handed out Grayson's motorcycle helmet. Not even a new scratch, Mister Grayson. The thief left it in the parking lot."

Bucky said, "If he had been wearing a helmet he might not have broken his neck."

Grayson puffed through his lips. "You know that's a crock, Bucky. These brain buckets cause as many neck injuries as they save skull crackings." He waved the helmet threateningly. "I wear this head gear under protest."

Leaning in the window, Grayson changed the subject. "Who were those two lumps in the pickup, Bucky? You know them?"

"Not really, but they're local. Names are Dillard and Crowder. No sheet on either man. They're in the County Militia, Bill said, and they make a lot of noise in a couple of different bars. They're just big and dumb."

"They're also criminals, Buck. They were in on the bike stealing, and we both know it."

"You're right, and we'll pass the word around. I can't get over that little weasel picking your motorcycle off right in front of two cops."

"More macho than smarts, Bucky, but he damned near made it. I figure he had to know exactly where he was going, and it had to be fairly close. He would not have wanted to be on the road long, especially without a helmet. He could have run onto a cop car almost anywhere. Nope, he was only going a little way, and that tells me that he had local connections, even if he was not a local guy."

Driving away, Bucky thought about Grayson and the accident. In his younger years, Grayson must have been a fireball. Hell, he rolled unharmed from an accident where the other guy got killed.

Bucky considered that a little. Weird that the thief broke his neck in a tip over at barely trotting speed. Maybe the guy had weak bones or something. More likely, the thief just landed wrong. At least it was quick and final for him. A lot of broken-necked people were paralyzed but lived on and on, their lives depleted and possibly miserable. Bucky shuddered. From where he sat, dying was better.

Robert Grayson thought little about the dead thief. If he was experienced, and if he looked closely, the medical examiner might wonder how the biker could have wrenched his spine so severely in both directions, but he would think *motorcycle accident*, where trauma could be severe, and move to the next body.

Gray recognized a small satisfaction that old skills remained, and that the thief would not be around to steal again. Otherwise, as usual, he felt nothing more.

4

Korea, 1953

Private First Class Robert Grayson steadied down behind his telescopic sight. The 2 1/2 power magnification was not much, but he was shooting at almost nine hundred yards, a nearly impossible distance without a scope.

He had sandbagged and camouflaged his foxhole so that it was barely visible from the front. When he fired, Grayson chose shots that angled along the line. His shooting was longer, but he never exposed himself or his position to the enemy.

The young soldier's rifle was a bolt action Springfield, Model 1903A4. Every US Army Company in Korea was issued the scoped Springfields, but most of the rifles were racked unused and unwanted in company supply rooms. American soldiers wanted their fast shooting M1 Garands that could pump seven shots at their enemy while a bolt action rifleman was getting off his second round.

Grayson saw it differently. He had asked for an 03A4. He had zeroed carefully and had proven that he could, as expected, be far more accurate with his bolt action than he could with an M1. When Grayson squeezed his trigger, he killed.

Because the Korean conflict was winding down, enemy attacks were usually small and of short duration. Hills were fought over and weaknesses probed, but the human wave assaults by Chinese infantry had ceased. Negotiators postured and line soldiers fought only sporadically.

PFC Grayson kept on killing. Within the protection of his fighting hole Grayson glassed his enemy, and a careless Chinese within a half mile was in danger.

Few appreciated Grayson's perseverance. The front was mostly quiet. Troops liked it that way, and officers preferred that enemy counter fire land elsewhere. To a degree, Grayson obliged. When an enemy fell from Grayson's bullet, the angry return fire was almost certainly directed at the Americans or south Koreans directly opposite the victim, and Gray's outfit remained unscathed.

The Company First Sergeant had come up to watch Grayson shoot. There had again been agitating to have the Sniping stopped. The Top Sergeant decided to see for himself.

"What have you got, Grayson?"

"Not sure, Top. Take a look with your binoculars. I saw strange uniforms, but the scope isn't strong enough to tell what they are."

The First Sergeant located Grayson's uniforms, three figures behind frontal protection but exposed from Grayson's angle. Doing some glassing themselves it looked like. Awful far though, hard to see. Sunlight broke through highlighting the scene.

"Holy hell! That's a Russian!" In his excitement the Sergeant's binoculars wandered. He got back on target.

"Which one?" Grayson's voice was unemotional.

"Middle one. Good God, he's got gold on his hat and medals on his chest. I can see them glinting. Can you . . ." Grayson's rifle cracked cutting short the Sergeant's question.

Jerked off the target, the First Sergeant fought himself back into focus "You tagged him, Grayson. By God, you dropped him. They're hanging all over him, and . . ." Grayson's rifle again startled him off balance.

"Damn," the Top heard Grayson's bolt working for a third attempt. Too late, the figures were under cover.

"I couldn't see the second shot go in, but you got the Russian with the first."

Grayson had his own binoculars up. "If I hit with the first, I got somebody with the second, Sarge. I put it right into the middle of them."

"Well, you got a God damned Russian, Grayson, I'll tell you that for sure. Whew, that was a hell of a shot. How far do you figure?"

"That bunker is eight hundred and twenty-five yards, Top." Grayson pointed out the target on his range card. I've test fired it when no one was showing. Once in a while I've had glimpses, but no targets there, till now."

"Holy hell, a Russian. A big officer with all of his medals, as sure as hell is hot."

Grayson said, "You could warn the outfit across from that bunker to hunker down because they might get some counter fire."

"You're right. Nice shooting." The First Sergeant went for a radio.

First Sergeant Ames was a Retread, a World War II soldier who had been called up from the reserve. In his civilian life Ames worked for the government. His primary duty was checking the backgrounds of individuals seeking secret and above security clearances. The position occasionally required a special skill, the ability to know if an individual fit a specific pattern desirable for a particular assignment.

Ames showed exceptional skill at judging special abilities or detecting often almost imperceptible inadequacies. Ames could pick the right man, and he could reject prospects with equal exactitude.

First Sergeant Ames knew a lot about PFC Robert, no middle initial, Grayson. PFC Grayson had caught Ames' attention when he had handled PFC Player with calculated dispatch. Grayson had been efficient and thorough but not heavy-handed, all attributes Ames looked for in many of his candidates.

PFC Jim Player was a snaky looking, mean eyed hillbilly from a nameless hollow in the Tennessee mountains. Player claimed to be a knife fighter and showed scars to prove it. With a few beers fogging already shaky restraints, Player could be abusive.

First Sergeant Ames had arrived late for the Player-Grayson fight, and the action had begun.

Background on the incident came out during his investigation. Player had verbally needled Grayson with little response. Frustrated, he showed a jackknife and suggested he might cut PFC Grayson for the fun of it.

Ames came in time to see Grayson make his move. When he started for Player, Ames was able to look straight into Grayson's eyes. Ames knew the look: calm, reasoned, confident determination. Genuine hard men had eyes like Grayson's.

Grayson drove into Player slamming him backward into a wall. Player's enraged curses evolved through a frightened squall and into an agonized scream. It took the same extended moment for watchers to realize that Grayson had Player's private parts in his grip and seemed intent on ripping them from his victim's body. For another long instant Player thrashed futilely, beating on Grayson's back, then he lost ability to resist and collapsed barely conscious.

Grayson let him fall, then stepped away. Most watched Player fetally huddled with his hands clutching his savaged parts. First Sergeant Ames kept his eyes on Grayson.

Interesting. Grayson's chest did not heave from adrenaline rush, and his eyes were still controlled, angerless, and gloatless. Ames was about to step in when Grayson seized Player's barely resisting right hand, gripped the two major fingers and broke them backward. The fingers dislocated at their bases with audible pops. Player screamed and lost consciousness. Ames felt his hackles rise and others in the crowd squalled in protest.

Grayson dropped the broken hand and turned to the howlers. His voice was dispassionate, as cool as if announcing the weather.

"That will keep him from coming at me with his knife." There was shuffling by some who were unsure about fair play, but they knew Player and recognized that Grayson was right.

First Sergeant Ames announced his presence, had Player sent to the medics, interrogated PFC Grayson and a few witnesses, and laid down his law. If any of this happened again they would all face court martial. As it was, both Player and Grayson could expect to remain PFCs for a long time, and neither would have a pass to go anywhere for damned near as long. Player's broken fingers healed slowly, and the man was transferred to a rear echelon unit.

Back on the line, Grayson again came to Ames' attention when a small firefight erupted in front of Grayson's position. Enemy sappers had gotten close. At early daylight they had blown the protective wire and charged through the gap. Firing exploded all along the front as American troops shot wildly at nothing and everything.

Grayson's platoon leader described how right in front of him Grayson had fired an M1 rifle seven times with careful deliberation. Enemy went down, and the attackers fled. Within the barrage of rifle and machine-gun fire there could be no confirmation, but the lieutenant believed Grayson had killed seven Chinese with seven shots. A commendation was discussed, but the lieutenant rotated home, and only First Sergeant Ames remembered.

There had been an intense and lengthy artillery exchange after PFC Grayson killed the Russian. The enemy concentrated on the unit directly opposite the Russian's position. American counterfire went after the Chinese and North Korean artillery. The same day First Sergeant Ames wrote his letter.

Dear Brad,
In September, a Robert Grayson, RA 11133168 will rotate from Korea. You should take a look at him.

Ames

Two months before his year in Korea was finished, PFC Grayson began to be promoted. Due to attrition, through sickness, combat, and the "serve a year and go home" rotation program it was not unusual for infantry soldiers to rise quickly through noncommissioned ranks. The most notable example was a Sergeant Haley, who after a total of nine months in service, was promoted to Master Sergeant. Haley was almost immediately transferred to Japan where he was both the envy and the rage of soldiers struggling to advance through more normal promotion criteria.

PFC Grayson skipped corporal and was appointed staff sergeant. A month later he made sergeant first class and served as Platoon Sergeant.

Grayson left his unit with First Sergeant Ames' advice to make something of himself.

SFC Grayson arrived in San Francisco after more than a month delay. A rear echelon clerk's helper, a PFC Jim Player, had "lost" SFC Grayson's shipping orders and finally managed to place Grayson on the slowest and crankiest troop ship.

Player did not make himself known until Sergeant Grayson was boarding. Then he positioned himself where Grayson could see him, smiled, and gave his enemy the extended finger. Grayson's cold eyes stayed on him too long, and Player grew uncomfortable. He turned and lost himself in the crowd, but Grayson-s dead-eyed recognition stayed with him longer than Player would have expected. Jim Player mixed in bad company, and he knew unforgiving hardness when he saw it. Player wished he had noticed before. He was suddenly glad that Robert Grayson was gone for good.

After processing at Camp Stoneman in California, returning Korean veterans were usually transferred to centers near their homes for discharge or further military service. SFC Grayson was pulled aside for special counseling.

That he reported to a civilian did not astonish Sergeant Grayson. He supposed that the army utilized non-military people to assist in most processing. The Korean Conflict was not a popular war, and the military's personnel sections were often underfunded and therefore understaffed. Healthy soldiers were needed in the war theater, not in California. Civilians hired for temporary work proved a workable solution.

Grayson was mildly surprised by his interrogator's directions. There was discussion of his boyhood along the rivers. The civilian, who wished to be called Brad, knew something about the juncture of the Juniata and the Susquehanna. He spoke easily of Perry County and the village of

Duncannon. Brad wished to know of Grayson's relationship with the grandparents who had raised him, and if he had surviving relatives, even distant ones. There were none.

Brad asked if he could call Sergeant Grayson by his nickname, Gray. Why not; Grayson preferred the shortened name. Brad, Gray figured, was a shrink, probably spot-checking combat men, part of a study maybe to determine if they were still rational.

Brad did know things. He knew about Gray's Sniping, and he added information. Grayson's Russian had been a General Officer who, according to *Investia* the quasi-official Soviet newspaper, suffered a motor vehicle accident. Brad wondered how Grayson felt about that.

The psychiatrist, if that was what he was, even knew about his fight with Jim Player. That got more of Grayson's attention, because, as far as he had known, Player's injuries had gone down as an accident involving falling onto and then off of a narrow beam. Grayson guessed that he was not a random selection after all.

Gray's troop shipment departed, and he had a barracks to himself for a day before another ship dumped its thankful horde. Sergeant Grayson did not see much of his new barracks mates. He underwent an especially detailed physical examination and talked through the other hours with his new friend, Brad.

Grayson was asked about his intentions. Gray thought he might stick with the army.

How, Brad wondered, would Gray feel about going back to Korea if the war hung on?

A good place to go, Gray explained. He could make Master Sergeant and perhaps move into a First Sergeant's slot. Grayson did not think beyond there.

But what did he think about killing? Did he hate it or have bad dreams?

No, it was just a duty.

Did he like killing people--maybe just a little bit?

Gray gave such questions serious consideration, but no, killing was just another duty.

Did shooting the enemy fit with his religion?

Grayson had no special religious convictions.

A God?

Possible.

Did God have special requirements that humans had to meet? Was he up there judging men who shot other men?

No. If there was a God he had bigger duties than watching human activities.

How did killing a man feel? Brad had never experienced it.

Gray considered and decided that he did not feel anything other than satisfaction with the shot.

How many men had he killed?

Gray was surprised by the question, but supposed it was natural curiosity.

Gray had never kept track. He took a long minute to conduct a rough count. Probably thirty or forty.

Brad was astounded and commented that most infantrymen never killed anyone. How . . .?

Gray explained about the Sniper rifle and noted that most men in a firefight lost control and shot wildly or in some cases did not shoot at all. He seemed unsurprised that he remained unaffected by the closeness and probability of death. That was just the way he was.

That was the way he was, and that ability might be useful--someday. Grayson's IQ was 132, unusually high, a first-class mind. His aptitude tests were also outstanding. Grayson was physically sound. His education was typical high school graduate which meant, not much. But, an empty vessel could sometimes be best. You did not have to spend valuable time emptying the useless and incorrect.

Brad, his full name was Bradford P. Gilmore and he was not a psychiatrist, believed Robert (NMI) Grayson had prospects.

Of course, it was hard to tell how a boy-soldier would develop, or, for that matter, whether the country five or so years down the road would need him anyway.

Gilmore's objectives were long range and at best barely defined. Gilmore's recruiting had to be speculative, but he also had to proceed now to be ready then, if the need remained. A requirement for the right men existed in the 1953 world, and the company and the country made do with whomever they could find, but the future looked stormy, and Brad Gilmore's rather narrow objectives emphasized having the right men on the payroll trained and ready.

Brad said, "I've got a better offer for you than the army, Gray. The pay will be better, the retirement is as good, and you will spend most of your time improving yourself including a lot of schooling and world travel.

"Does that have appeal?"

Grayson's eyes held Gilmore's until Brad's slid away.

Finally Grayson said, "You want me to kill people, don't you, Brad."

Grayson's words were not a question, and Gilmore was jarred from his comfortable routine. He heard himself stammer, "I haven't even suggested such a thing."

Grayson's direct gaze brought him up, and Gilmore dragged his thoughts together. How in hell had Grayson deduced that? The young soldier's observation certainly cleared the air--only it was a year or more sooner than Gilmore had intended.

"How would you feel about that, Gray?"

Grayson's reply was as calm as any of his others. "I would have to believe it was the right thing to do."

"That's all? If you believed it right to kill someone you did not know you could just go and do it?"

Grayson showed mild surprise. "If that was my duty, of course I would do it."

"Well, mightn't you go out and kill someone on your own without orders, just because you thought they deserved it?"

Grayson frowned in thought. "Everyone wants to kill someone once in a while. They don't do it. I guess I am like everybody else."

Well, Grayson was not like everyone else. Brad already knew that. He went ahead.

"So, are you interested, Gray?"

"I'm interested, Brad."

5

Robert Grayson came home from the war and a local man asked, "How are you, Robert? I haven't seen you lately."

Gray said, "I'm fine, Mister Cole. I've been out of town for a while."

"Oh?" Complete disinterest. "Well, nice to see you."

So much, Gray thought, for the war hero stuff.

Gray found former friends busy with their lives. Most had married, usually to local girls. Often the couplings seemed improbable, knowing what all knew about each other.

There were few questions about his war experiences, even from others returned from Korea. Life went on, and most put their pasts, whether social or military behind them.

To inquiries, Gray announced that he would work for the federal government. Where and in what capacity was still not certain, but he had a job promise. No one asked more.

Gray lived in a room in the hotel on the town square. He walked by his boyhood home, but the place was changed with fresh paint and a new fence. A noisy dog discouraged loitering.

He hiked uphill and along the ridge to where the road ended. Woods began there, and he could walk through until he came out in his favorite rabbit hunting field. The field was no longer farmed, and the briar patches housed a large rabbit population. The river rustled just beyond a tree fringe where the drop had been too steep to cultivate. Once, he and a buddy long moved away had cooked hot dogs on sticks over their smoky fire and kept an eye on fishing lines tied to whippy tree branches.

The place smelled and felt good, and Gray thought he would like to build a cabin in the trees just above the river. When his work allowed, or if life seemed heavy, he could escape to this, his boyhood campground, sit on a rickety porch and listen to traffic across the river.

If he got money, Gray decided, that would be just what he would do. The land was sitting there unused. Sooner or later it would come up for sale. The trick would be to know the owner and make sure he would call or write when he wished to sell. Gray hoped it would not be too soon. His savings from military pay were puny, and he doubted his government position, whatever it would be, would bring in any big money.

What a strange job offer he had accepted. Brad had promised schooling, travel, security, and work that he could handle.

Doing what? Actually killing people, as Brad's questioning had implied? The interrogator had neither confirmed nor denied, but Gray's question had startled and intrigued him.

Even though he had asked it, the question made no real sense to Grayson. He could shoot better than most, but shiploads of men coming home could do that. Why pick Robert Grayson out of the mob? Gray had asked, and Brad had said that First Sergeant Ames had recommended him.

"Recommended me for what?"

"Recommended you for consideration." Brad had answered.

"Consideration for what?"

"To work with me."

"Doing what?"

"Whatever I ask you to do."

Questioning Brad was about as effective as shoveling smoke. All Gray discovered was that he could have a job, and how well he worked out would determine how far he would go.

Gray liked it. If they stuck him behind a typewriter or between filing cabinets he would bail out, but as he had no clerical experience that would not be his kind of work.

What had he done that First Sergeant Ames could have noticed? He had shot the Russian, but how that qualified him for attention, Grayson could not imagine.

+++

Robert Grayson did not even have an official desk. He received directions via telephone or letter. His government supplied apartment was in northeast Washington. He was required to check in at a local number morning and evening unless told otherwise. If there were instructions, he carried them out.

He had a bank account. A biweekly deposit appeared there in his name. The amounts were satisfying. When he asked, Gray was informed that his deposits came from the Smithfield Corporation. Gray was not surprised that there was no listing for that company.

Gray picked up and delivered to peculiar addresses and locations. He became familiar with government offices and back street shops. When he saved enough money he bought an old but dependable automobile and became more efficient.

He was ordered to the FBI Academy to attend a basic agent's course. Well now, was he to be a "G" man? Perhaps not; at the academy he was listed as Simon Brown.

Simon Brown did well but never again appeared on government rolls.

Robert Grayson was enrolled at the University of Maryland's College Park campus. His courses were typical freshman but included no sciences. Being paid to go to college was comfortable, although Grayson's deliveries often interfered. By his sophomore year Gray was flying about the hemisphere, dropping in here and visiting there.

Gray saw Brad irregularly, but he met no other superiors or colleagues.

Gray asked, "Am I really working for the United States Government?"

Brad appeared surprised. "Of course. Your name is listed with the Congresional Budget Office. You hand carry sensitive materials for them."

Gray's brows rose in doubt. Brad shrugged, "Check with government listing. You are there. Your social security is paid, and you are in the federal employees retirement system."

"So, what am I preparing for?"

Brad smiled, "Your work will be varied."

"Still not telling me, huh?"

"We'll go into detail later.

"By the way, you are scheduled to go to Camp Stone Bay in North Carolina. You will be run through all of the Marine marksmanship courses. Your name will be Arthur Rubin."

"I've already been through everything that the FBI and Quantico have."

"Good, we will expect you to do well, and we are just beginning. Oh yes, you are to attend a Dale Carnegie personality development course, and you will sign yourself up with the Arthur Murray Dance Studio just down the street. Use your own name for these. Learn all of the dances, and I mean everything. You might need them."

"I'm trying to go to college, for God's sake."

"And you are doing well. Keep at it."

On his way out, Brad added, "Oh, one other thing, Gray. Do not become seriously involved with a woman. Have all of the girl friends you desire, but remain free of entanglements, or we will drop you like we would a hot coal."

The coordinator was gone before Gray could even sputter.

+++

In 1959, Robert Grayson graduated with an undistinguished B.A. in Sociology with minors in German and French. He missed his graduation ceremonies because he was in California attending a new kind of hand to hand fighting course.

Oriental martial arts were gaining in popularity, and Gray was instructed to become proficient in various schools and to thoroughly understand the programs.

Gray spent a year at it. The training was exacting, and for once he was not diverted to other tasks.

Grayson admired the city along the magnificent bay, and California itself possessed a dream-like ambiance of *try anything* with neither success or failure as important as the experience of experiencing.

Brad arrived without warning, waved Gray from his physical drill, and they stepped outside to a small Japanese style garden where their talk would be undisturbed.

The handler liked what he saw. At twenty-three, Grayson was lean and ropy muscled. He stood erect and appeared taller than his almost six feet. Grayson still wore his hair in a military flat top, but his next assignment would change that.

The preliminaries were over. Grayson was now established as a longtime clerk in Budget. He was an undistinguished college graduate. His bank saving and checking accounts showed a steady but ordinary increase with no question-raising fluctuations.

Grayson was traveled. Clerks in embassies knew him as an experienced courier. Grayson could come and go without particular notice, and his uninteresting record of personal achievement could stand any investigation likely to be conducted. As a courier, Grayson had a Top Secret clearance, but so did anyone handling sensitive materials. The mandatory FBI background check further supported Robert Grayson as an innocuous, lower level federal clerical employee unlikely to rise far beyond his current rating. Of course, Grayson's dossier did not include the many special skill schools.

Brad asked, "What have you learned, Gray?" The years had encouraged a limited informality between them, although the relationship was stunningly lopsided. Brad knew everything about Grayson. Gray knew only Brad's name; if the one he used even was his genuine name.

Gray answered, "I have learned a lot. These oriental skills can be powerful weapons. The philosophies taught are intended to curb and direct the skills' violences," Gray shrugged, "but few continue training intensively enough to become masters anyway."

"If you met an opponent, an enemy, who knew these skills could you defeat him?"

Gray's answer was as direct as if he had expected the question and had studied his response.

"In my opinion, Oriental fighting is not particularly effective. It is hell on wheels against an untrained man, of course, but otherwise," again Grayson shrugged, "it is not that good."

"Really?" Brad was surprised. "We've heard these arts, karate and the rest, are deadly beyond belief, and the demonstrations we've seen are damned impressive.

"Uh huh, but here is how I see it."

Gray paused to line up his argument. His opinions flew in the face of popular belief. A karate fighter of any of a half dozen school, it was believed, could slaughter an individual untrained in the art.

"First, there are few halfway measures in karate. It is not a skill useful in subduing an angered innocent. You are too likely to injure severely.

"We have had judo for years and years, so I do not include it in the new fighting we are talking about. With judo, you break holds applied to you and use other holds and throws to win. In essence, you get hold of your opponent and never let go.

"Next, unlike a plain old fistfighter or a judo artist, a karate fighter has to stay in practice. His moves are dramatic and exacting. If you do not practice you may not have the balance when you need it.

"Now, this is the heart of it. Karate fighters like distance. They need space to perform. A karate fighter usually cocks back on his rear leg. His balance must be rearward to free the other leg for kicking. They almost invariably begin with their feet or a series of supposedly devastating arm punches.

"You fight a karate man by swarming him. Charge and grab him. Smother his stuff. Use judo and plain old gouging, kicking, biting, and slugging street fighting. Jam him, crush his testicles, clamp your teeth on his nose and try to chew it off. Break fingers and rip his hair out."

Gray breathed deeply. "Kill the son of a bitch right then and there."

Brad felt his neck hairs tingling, and Gray's follow up did not settle them.

"All of this assumes you have not had a chance to shoot him, because a martial arts expert is always dangerous, and a bullet in the chest ends it for sure."

Over lunch Brad said, "We have another school for you, Gray. This one teaches a lot of Sherlock Holmes kind of stuff. You learn really advanced tailing techniques, how to apply and use disguises, a lot on silent weapons, that sort of thing."

"Sounds interesting."

Gray frowned and his eyes pinned Brad's.

"I think it is time we talked in plain language, Brad. I have been training for five years, or six if you count Korea. What is the goal? Spell it out because I need to know. I'm full grown and college educated. I am not the boy you started with back in '54. Lay it out now, Brad.

Brad Gilmore nodded acceptance. It really was time. Of the few individuals he ran, only Grayson was measuring up.

That was all right; it took years to be sure about a man. It took years and vast expense to guide and develop what Grayson would soon become. To speak too soon could scare off a prospect, but to wait too long could be to lose to other interests. Women were the worst danger, but in Grayson's case only one female had become a potential threat. That woman had unexpectedly received a too good to refuse job offer far away, and Robert Grayson had remained on course.

Brad studied their surroundings. The restaurant was clear, a Chinese place with a Japanese garden to favor Grayson's recently developed taste for rice and chicken with a lot of broccoli mixed in.

Finally ready, Brad was direct.

"Do you know the term stone killer, Gray?"

"No, it's new to me."

"The Mafia likes the title. If the mob requires an elimination that must not be traced to them, they arrange for a stone killer.

"The idea is to bring in someone from so far outside that there is no connection to the organization. The assassin never sees his employer. Individuals called cut outs, who themselves only hear telephone voices, are used. The stone killer makes his hit and is gone. He is paid by the cut out who is known to the killer. Not to pay could cost the cut out his life, so the cut out has his money in hand and ready before the stone killer strikes.

"Assuming the assassin does his part, police, FBI or the mob are unable to pin anything on anyone."

Grayson nodded understanding, a small cynicism showing at a mouth corner.

Brad went on. "Governments too need such men. Not often, but when needed, the task can sometimes be beyond the skills of a gangster type stone killer.

"The Mafia often kills by importing some desperately poor peasant from Italy who can handle a shotgun or a pistol. The destitute family receives up front money with the promise of more when their man succeeds.

"The poor slob is brought to the scene, makes the kill, and is sometimes executed himself and buried in the Jersey Pine Barrens. No witnesses is important. The family is always paid, with a bonus and an explanation for the unfortunate death so that other killers will be available if needed.

"Clearly, that kind of smash and grab is not sustainable by serious governments. A government must succeed without clumsy involvements and certainly without all of the inherent treachery and dangers of eliminating their own men."

Brad paused, allowing Grayson time to speak, but the younger man merely waited.

Gilmore was comfortable with his prospect's calm patience. It was, in fact, expected. At least in part, Grayson had been chosen because of his ability to remain calm and unemotional.

The handler went on. "Of course, our government does not engage in assassination. Any member of Congress would rail against it. The head of any agency would be revolted by the thought. We have laws against assassination, we speak at the United Nations against it, and we publicly abhor such immoral activity. Other countries may succumb to temptations and eliminate enemies or competitors but never our country."

Gilmore's smile was cold. "Keep those facts clearly in mind as we speak."

They walked a little before Gilmore spoke again. "In theory, under the best of systems, a killer for a government should be a highly trained individual with an impenetrable cover. Some are beginning to refer to deeply buried agents as Moles, but Moles usually live in countries other than their own, for obvious reasons. Moles are in place for years and surface only for special uses.

Moles, whether in or out of country, have genuine jobs, families, and communities. They become involved in all of the normal activities, PTA through church service. Their normalcy is their protection, and Moles are called upon with great reluctance because once detected, the huge investment is lost.

"So, if a government believed it might have a future need for, for instance, a trained assassin, it might choose likely individuals and bring them along over a period of years, carefully observing their development, yet revealing nothing until a prospect's progress indicated the time was right."

They stared together into the tiny pond that highlighted the garden. In the Oriental manner each stone and flower was exactingly placed to enhance the sense of natural tranquillity. A strange place to be speaking of killing and secretive maneuverings, Grayson thought.

But, perhaps not. There were parallels between the developed and labored-over perfection of the pool and the lengthy training and disguising of a perfect assassin.

In the mutual silence, Gray found his awareness expanding. He supposed he had always known it was to be something like this. The endless training with weapons, the travel and resultant familiarity with distant places were more than clues. He had been educated and trained. He lived and moved easily in a capital city, yet he also hunted, fished, and was natural with rural people. He could blend within many societies.

But an assassin, a killer of anyone? That concept was more than a simple leap forward.

Gilmore was ready to talk again. "What a country might do would be to recruit young men who possessed special attributes that might indicate not only ability but a willingness to serve their country in a violent and never to be recognized manner."

Grayson finally chose to interrupt. "But, why in hell would anyone pick me, Brad? I was barely eighteen years old when I went to Korea. What attributes did I show, for God's sake? Shooting that Russian General? Everybody knew that just happened; it could just as well have been a Russian Corporal for all I knew or cared."

Gilmore appeared surprised by the question and chose to abandon the artificial supposing.

"It was not the General, Gray, although he was a sort of clincher for First Sergeant Ames.

"Didn't you wonder just a little what your First Sergeant was doing up on the line watching a Sniper work? Ames was observing you, not your shooting."

Gilmore sighed, "Ames had an eye on you ever since you fought Private Jim Player. He . . ."

"Player?" Grayson was astonished. "Good God, I'd almost forgotten him."

Gilmore's head nodded, "Ames decided that you demonstrated a certain, ah . . . restraint and quick decision-making that was unusual."

Grayson tried to remember. "Hell, I broke two of his fingers so that he could not use his knife on me. That didn't take a lot of reasoning."

"Ames' report also remarked on how unaffected you were by killing and death in general."

Grayson thought about that as they chose seats beneath a brightly painted parasol.

Gilmore asked, "Is this garden Chinese or Japanese?"

"Sort of mixed." Gray wished to stay on their subject.

Gilmore said, "You once asked me if I wanted you to kill people."

"I remember."

"I asked how you would feel about that."

"I remember it all."

"So, how do you feel about it now, Gray? It has been five years, and you have grown. Once you killed without emotion, and a successful killer must do that."

Grayson answered easily. "I have heard about soldiers who have bad nightmares about war and killing. I know a cop who shot one man and is not sure he could shoot again." Gray shook his head. "I share none of that."

He leaned closer to Gilmore, as if to impart a secret. "Brad, I cannot tell you why, but it does not bother me even the tiniest to kill a human that I wish to kill."

Gray again straightened. "If you asked me to shoot dead the next man we happened upon, I would not do it, and I would feel rotten if I did, but kill an enemy? Without a blink!"

"On this job you would never really know your target, Gray. You would know about him, of course, and he might be a famous name. You would receive the assignment through me and be expected to carry it out."

"I would have to decide for myself each time, Brad. I would not kill just because you said someone else said to."

"You mean that you would have to know why you were to kill?"

"Exactly, and to approve."

Gilmore was curious and chose to be personal. "How do you feel about humans, Gray? I know I have asked you that before, but how would you square killing with God, for instance? Wouldn't you think or care about grieving widows or orphaned children?"

Grayson was just as serious, but he had to frown while working out his words.

"In war I killed enemy soldiers in piles. They probably had families. How were those men different from any other individuals I might be asked to kill?"

Gray shrugged with some bafflement. "I know that when they killed, other guys felt something that I did not.

"I know that most soldiers were scared to death, so badly scared they could barely function. I never felt that. I know that I am different somehow, and First Sergeant Ames must have seen that--whatever it is."

He again leaned closer. "I could kill a man for kicking a dog as quickly and with as little regret as I could kill the dog for attacking a man. I am absolutely certain of that, Brad."

Gilmore studied Grayson's eyes for a fanatic's glare or the dull stare of the unaware, but his prospect's eyes were as usual almost boyishly alive and neither charged nor secretive.

"If you are right, you have a rare talent, Gray." Gilmore was sure of that. Under stress, a typical man's blood pressure soared and he acted less rationally. Grayson, all of his testing showed, might be different.

A later First Sergeant Ames letter had included,

> *"Robert Grayson appears to be completely unaffected by conditions that would panic or paralyze most men.*
>
> *He remains normal in actions and appearance in the face of personal danger and in the confusion of infantry combat.*
>
> *I have to tell you, Brad, Grayson is calm without being cold. He reasons when most would have simply reacted. I saw him whip the hell out of a mean-assed knife fighter. Grayson squeezed the guy's gonads into mush. That was impressive, but he let go without killing the guy and did not kick him in the head as we might have expected. Instead he broke the man's fingers so that he could not use his knife. Amazing, Brad, no excitement, no sweat, I doubt his blood*

pressure changed from 120 over 70.

Grayson kills the same way, business-like, no nonsense, and absolutely no concern over the killing. I've watched, and I've gotten other opinions. He could be a REAL prospect for any sensitive work you might have . . .

Gilmore got himself back to the present.

"So, Gray, you should think about our task. Continue in your training, and maybe sometime soon I'll have a proposal for you to consider."

Grayson thought about Gilmore's explanation. If he were not inwardly certain of what Gilmore had in mind, he could not be sure that he was actually being groomed as a government killer. Gilmore had talked all around the subject and had asked his opinion, but nothing definite had been offered or accepted.

Gilmore would be back, but probably not before the Sherlock Holmes school was completed. Gilmore would have an assignment, perhaps to kill instead of carry.

Gray wondered what he would do. He really had no objection, as long as he was convinced the killing was worthwhile.

Worthwhile? Grayson had to smile over that. Damned few in this world would agree that any assassination was worth anything. Oh, maybe a Hitler would be all right, and deep inside a lot of people might approve of someone unknown knocking off an infamous gangster or an opposing political party boss, or even the town bully, but they would never admit to it.

Ames had apparently thought he had a talent for killing, and Gilmore clearly believed the same.

Interesting, Gray would think about it.

6

Gray celebrated his twenty-fourth birthday by buying the unused farm above the river. His pay had again jumped, and he could manage the bank payments. A man who did not drink or smoke usually found money accumulating.

He was not astonished when Gilmore called to congratulate him on the purchase. Owning land, Gray gathered, enhanced his for-real status in the world.

As to Gilmore's knowing? Grayson was not surprised. Gilmore knew everything. Gray handled that by keeping all that he did correct and aboveboard. No secrets--no problems, but Gray regularly searched his apartment for bugs. He had completed training on bugging and debugging and knew what to look for. Spying on his home life would have been too close to be tolerated, but Gray found nothing and believed Gilmore was too wise to intrude.

Good old Brad had a lot of years invested in Robert Grayson, and Gray doubted that Gilmore would pointlessly endanger his investment.

The farm became an important escape, and Gray often drove the four hours north to spend an afternoon on his land.

As a new owner, Gray's first impulse was to post his one hundred plus acres. Hunters would be turned away and no trespassing signs would discourage other intruders.

Bad idea! Generations of local men had roamed the fields and woods of Gray's farm. Undoubtedly, there were boys much like he had been who camped and shot at rabbits on his new property. So what? The land was rich and game was not hunted out. He had no legitimate reason to close the land. In fact, he could recall old timers during his youth who had complained about stingy absentee owners whose first act was to post their land against hunting.

Gray recognized with some irony that posting would not amount to much unless he was present to enforce it. Local men knew that deer sought the safety of closed-to-hunting ground. The locals would organize and drive posted land whether the owners liked it or not. Hell, deer were public property, and that was the American way.

Gray opened himself a shooting lane that ended against a steep embankment, and rare was the visit that he did not fire his weapons. Gray picked up his empties. If he had not, an observer might have wondered about who shot away so many rounds of pistol and rifle ammunition.

Gray drew his ammunition from the Quantico Marine Base and the FBI. His FBI ammunition was delivered to him. Gray had not approached the Hoover building or its people since his beginner classes there.

Nor did Grayson appear at the CIA's Langley Headquarters. Gray's classes were conducted in rented buildings, and since his Sherlock Holmes course, Gray's instruction had been one on one. For classes he used only a false first name, and knew each instructor only by a similar name.

It was clear to Grayson that he was nearing the end of formal training. It should end, he believed. Increasingly often he already understood the instruction and only a little was new material.

Gray now had a cubicle among three dozen similar enclosures in an obscure government office building. Gray had a desk and a telephone.

Most days a stack of completed forms requiring Gray's signature were placed in his in-basket. He signed each one and moved the stack to his out-basket. Anyone investigating Robert Grayson would discover a body of actual work.

Gray spent little time at his desk. He jogged along the canal, and he fired on a number of different shooting ranges. He most often went to Quantico. The Marines were no nonsense, and they were accustomed to agents and the like who used civilian guns and practiced strange shooting positions.

Quantico also provided Gray with gym facilities. Grayson worked out diligently and played racquetball with military and other first name only acquaintances.

Still, there was a pointlessness to it. Peacetime soldiers suffered the same syndrome. They endlessly trained for what never came. After a while a war would almost have been welcome.

Gray's was an almost friendless life. He met people and enjoyed limited socializations, but when anyone drew too close, Gray eased himself away.

He was not sure why he avoided intimacy. Having close friends would not have exposed him. So far, in fact, he had nothing to expose. Almost weekly, sometimes more often, he was dispatched to a distant place with a message or a parcel, but that was a courier's job. Gray Grayson? Everybody everywhere seemed to have encountered him. Grayson's cover was to be like an old shoe, comfortable, familiar, and unnoticed. Unnoteworthy, in fact. Who could be interested in the activities of a Grayson?

Brad Gilmore came in person to Gray's cubicle. He had a name and a task.

Gilmore said, "Jay P. Worthington. Eliminate him. Choose your own time, place, and manner. There is a dossier on him at the Smithsonian. Use it there only."

Gilmore raised his eyebrows inquisitively, "Any thoughts or questions, Gray?"

"Can't think of any."

Gilmore rose, "On the time part, Gray, take what you need, but sooner is better." The handler departed.

Gray sat for a while. Just like that everything had changed. He guessed he had expected more. A secured room, projected photos of the target, an unidentified over-voice monotoning, *"This is Jay P. Worthington, a murderous, evil creature wanted by all humanity. Your mission, if you choose to accept it . . ."* Gray hoped the dossier had good meaty information.

It did. Jay P. Worthington was a former American who lived in England and used a Swiss passport.

The dossier had a lot more. It was almost a library in itself. Worthington had attracted interest.

J. P. lived on a too large estate in Avon, against the sea. His wealth was great and had passed from father to son. The fortune had begun in the 1920s with a grandfather who dealt in slaves. Good God, slaves? Young women mostly, sold to Middle Eastern potentates.

The father's activities were poorly chronicled, but the money had piled up. During World War Two, Worthington, senior had transferred between Ireland and Switzerland avoiding unpleasant wartime duties.

J. P. Worthington had gained the family wealth when his father had inexplicably driven a car off the "Going To The Sun highway" in Glacier National Park. *Hmm.*

J. P. had surrendered his American passport and denied his citizenship. In so doing, he also avoided a monstrous tax bite. Still, all of that was more or less understood as high finance. Grayson still looked for the need to kill Jay P. Worthington.

The reasons became rather quickly defined. Worthington bought generously into smaller but prospering firms. Included in each investment was an aside that the business would, in the inevitability of eventual demise, belong in entirety to the survivor. *Hmm!*

In nine such American investments and in seven foreign purchases, Worthington's partners suffered almost immediate and peculiar demises. One actually "fell" into a zoo's lion pit. Three were simply murdered. The rest were suicides. One suicide, who left a signed note, had shot himself three times in the chest.

The law was helpless. Worthington was always far away and in the presence of many. But it had to stop.

Notifying companies of Worthington's suspected methods could not work. J. P. was already purchasing businesses through third parties, straw men who quickly sold their shares to him.

Robert Grayson was to earn his spurs by removing Jay P. Worthington in any way he chose.

Any way, of course, meant untraceable. There was also to be no martyrdom involved in Gray's work. Grayson was an investment not to be lost through idiocy. Gray began to look for the best way.

There were rich opportunities, and although Worthington kept bodyguards handy, he was a creature of habit who could be expected to be certain places at certain times.

Gray studied at length assembled testimonies of individuals who had, or in some instances still, worked for Worthington.

A short commentary caught Gray's attention.

". . . The Mister walks

most afternoons. He enjoys circumnavigating his lake, which gives him nearly a mile of exercise . . ."

Gray turned to aerial photos of the Worthington estate. From directly above, portions of a path around Worthington's lake were visible. An oblique photo showed the estate's traditional stone perimeter wall and betrayed the barbed wire topped fence further in. There were roving patrols mentioned but no details. Dogs? Perhaps not. It was noted that *The Mister* was allergic to animals.

Jay P. Worthington was an unprepossessing 5' 9" tall, running to fat, and was bullying in manner. Gray saw no difficulties in those details. When gun carrying was mentioned to various information sources there had been only amusement. Of course the bodyguards were armed, sometimes with billys, but occasionally with military pistols. *The Mister* relied upon them, and as far as was known, owned no firearms. Grayson, for the first time, approved of Jolly Old England's insistence on disarming its populace. Over there, the criminal could be quite certain of his victim's helplessness. Not so in the United States. Here an intruder had to expect facing a pistol or shotgun. Gray believed the presence of guns in honest hands made the bad guys more respectful. In Worthington's adopted country, Gray would be a bad guy. He would not chance bringing in a gun, and it was nice to know that his opponents would probably be equally ill-equipped.

The estate seemed the simplest point of attack. If he did it right, Gray figured he would be in and out undetected. Gray thought Gilmore had chosen an excellent final examination, well researched, with no time pressure, and a relatively soft target.

Gray studied Worthington's photos and thought about killing him. He felt nothing more than a tingle of comfortable challenge.

It was early summer, a best time for outdoor walking. If Worthington did walk, he would be out now. Gray had chosen the place, and he had decided on his method.

Gray memorized what he needed, returned the file and studied plane schedules out of BWI airport.

No substitute identification had been offered. That too was wise. He would simply be Robert Grayson, innocent American traveler, often in England, known to many. He had certainly performed that roll enough times.

Gray flew coach to Heathrow and trained to the seaside village of Clevedon. He took a room in a modest hotel overlooking the narrow beach line and sought a motorcycle rental.

Announcing his intent to explore the historic Avon, particularly medieval sites, Gray chose a nondescript BSA of 1970s vintage and putted about the streets gaining a feel for the machine and for the left hand traffic. The BSA started with single kick reliability. Its engine had worn to a smooth if smoky dependability. Gray was satisfied with his transportation.

The perverse English climate failed to provide comfortable walking weather, so Gray used the drizzly day to cycle past his selected entry point to Worthington's estate.

Traffic was light along the narrow ribbon of back road. He should have no trouble moving about undetected. Gray could see Worthington's stone wall, but the high fence was hidden by untended growth. Excellent, he would not be seen surmounting the barrier. Gray had no intention of cutting his way in. He would leave as few traces as possible.

The next two days were no better, but Gray used the mildly inclement conditions to establish himself as a legitimate tourist. He took notes and ran rolls of film through his camera. He discussed the excitement of his finds with any willing to listen and sought advice on how to copy gravestone markings by rubbing charcoal across paper placed against the stone. He made a few rubbings to have among his possessions if inquiry came his way.

The third morning promised clearing by noon and a bright and sunny afternoon. If Jay P. Worthington was in residence, as he almost always was, he might walk. Gray prepared himself.

Unlike Agent 007, Gray had no smuggled in Walther PPK or even a Commando's Fairbairn killing knife. Gray did have a sock with a small handful of coins knotted into the toe. If asked, the knot kept the handy coins secure and prevented their infernal clattering in his pocket. That the weighted sock also constituted an effective blackjack would be a surprise to him.

Gray wore inconspicuous dark clothing and had a fifteen foot length of one quarter inch braided nylon line stuffed into a pocket. The rope was English and of the most common type that was sold in a thousand stores a thousand times a day.

Before the weather cleared, Gray motored to his destination. He pushed the motorcycle into concealment within a thicket across the road from the estate. In the unlikely event of its being noticed, any connection to the Worthington estate would be obscured by being on the road's far side.

He crossed quickly with nothing in sight. He vaulted the stone wall and pushed through the wild growth to the fence. As expected, the barrier was a common hurricane weave, eight feet in height, with the favored three strands of barbed wire angled outward at the top.

Gray placed his gloved hands carefully, gripped the fence and listened intently. He heard only natural silence. He went up in a powerful surge, only his toes causing the fencing to chatter.

The barbed wire was tricky, and a climber needed to hold his body clear as he first hooked a toe and then muscled himself upward until he was poised horizontally above but not lying on the wire. The rest was easy, and Gray dropped into a silent crouch on the unfriendly side of the fence.

A narrow maintenance road ran along the inside, and it appeared well traveled. Gray stepped carefully, pausing to obliterate his barely detectable footprints.

Within the relative safety of Worthington's woods, Gray rested to test the wind and judge his emotions. The forest was silent. After a moment a clutch of motor vehicles passed on the public road.

Gray found his pulse at a high normal and only intense interest in the problem at hand touched his emotions. He considered that he hoped to kill a man within a few hours. He searched his mind but discovered no reservations or special tensions in the matter. His talent, Brad Gilmore had called it.

Baker Bill usually accompanied *The Mister* on his walks. Worthington was a stroller who paused to contemplate, a boss who often talked to himself and discouraged interruptions. Worthington carried a notebook and paused on occasion to jot reminders.

The walks bored Baker nearly to exasperation. If J. P. had stepped out and made an exercise out of them, the walks would have been different, but bodyguarding an employer who dawdled required exceptional patience, and Baker Bill, who had pounded out more than his share of pugilistic opoponents,, was not noted for patience.

At least they were out of the house. The custodial staff was a crusty bunch, each determinedly guarding his or her's space and privilege. Baker avoided them, but even the gymnasium and the elaborate telly equipment eventually grew stale and wearying.

For a while, Baker plodded behind his employer, wishing they would abandon the usual lake circuit and cut through the woods to view the sea. It was not to be. *The Mister* poked along, lost in thought, pausing to gaze unseeingly.

Bill wandered ahead and strode to the water end of a long boat dock. His employer still behind in the woods would join him. He always did, eventually.

The darkness behind J. P. Worthington shifted and closed in a soundless rush. Unsuspecting, Worthington pondered a clever maneuver when an arm slid beneath his chin and his neck was squeezed tightly between a forearm and biceps.

Through an instant's surprise the pressure of shut off carotid arteries was almost pleasant, but the need for air surfaced, and Worthington clawed frantically at the iron muscled arm.

Before asphyxiation could occur, Worthington's brain began shutting down. Only seconds of blood supply loss were needed to lose awareness. Only a few additional instants were necessary for permanent disability, and without freshly oxygenated blood, death followed quickly.

Worthington's hands fell away at the seven second count, and his body hung limp.

At fifteen seconds Grayson lowered what had been Jay P. Worthington and snapped his length of cord in a seaman's clove hitch around the neck he had squeezed. The clove hitch was a knot that tightened the more it was pulled.

Gray flipped the rope end over a branch, gripped Worthington by the belt and heaved upward. As the body rose, Gray cinched hard on the line. Worthington swayed, his feet still on the ground.

Gray heard no sound from the guard. He heaved again and tightened. Now the body swung free, its head grotesquely tilted from the line tightened to disappearance in the soft neck fat.

Grayson lashed the line's end to a nearby sapling and slid silently into the woods.

Baker Bill grew impatient. He hated to go back because then he would have to again walk to the dock's end with *The Mister* to stare silently at what had already bored him witless.

Baker retraced his steps, only fifty or so steps. The horror of Worthington swinging at a rope's end staggered him.

The bastard's killed himself was the bodyguard's first impression. He sprinted in and gripped the body, attempting to raise it enough to loosen the strangling noose.

No good, it was not working. Baker fumbled out his knife and sawed at the taut rope. The line parted, and the body fell like a meal sack.

The bodyguard struggled with the rope buried into the dead man's throat, but the line was deep into the flesh, and he could not figure out the knot.

In desperation, the guard cut everything near the back of Worthington's neck. He sliced through and got the line free. There was no bleeding. Jay P. Worthington was stone dead.

It was not a suicide. Baker Bill realized that even as he fought to free Worthington from the noose. There was nothing to stand on. The man could not have hung himself.

But how? Baker had been within easy hearing distance. Who? There was no trace. His own struggles had eliminated any earlier markings.

The guard peered around. He listened, but there was only Jay P. Worthington's terminal silence. Baker Bill ran for the house.

Without protruding barbed wire on the estate side, Gray flew over the fence. He vaulted the stone wall and checked the road for traffic. A vehicle had just passed, and Gray waited in hiding for it to move away.

He had crossed the fence at a different spot and had to trot a hundred yards to the hidden cycle. Gray wheeled the machine clear of brush and it started on the second kick, idling with its comfortable worn-in rumble. Gray settled into the saddle and putted away. A short minute gave him turns in the road, and he was soon motoring north on the major artery.

Except for concentrating on the still unnatural left hand driving, Gray had freedom to consider his act. Neat enough, he concluded, but far too much left to chance. There could have been a dog or an entire hound pack. Worthington might have walked with a group. Someone could have come along.

Gray was also displeased by his stringing up. No one would believe the death a suicide. A circus gymnast would have been hard put to hang himself in Worthington's manner. He should have squeezed J. P.'s neck for another thirty seconds and let it go at that. He had not because he wanted Brad Gilmore to know that Worthington had been done in. A corpse with its neck nearly parted by a thin cord was hard to declare accidental or by natural causes.

Grayson vowed to never again leave a deliberate trademark. If Gilmore considered Worthington a test, he had his results. Henceforth, Gray would do only what had to be done.

Regret, trembles, rapid pulse, disconnected thought--Grayson experienced none of them. He reviewed the scene as coolly as he would have a large meal. Gray's talent? Interesting.

Brad Gilmore listened carefully to Robert Grayson's description.

He said, "I thought you might choose a rifle or pistol."

Surprised, Gray answered, "I felt them unnecessary and dangerous to set up."

Gray was apologetic. "I should not have hung him up." His grin was boyish, and a matron at a distant table felt the impact. "That was for you, Brad. I wanted no doubts."

Gilmore harumphed, "There was no doubt, and you delivered a powerful message. No one has yet jumped to grab Worthington's reins."

The handler sipped at his coffee. "You were wise to continue your sightseeing for a few days. Planes and trains may have been watched, and the less your name shows up during crime investigations the better." He added, "It was a good job, Gray."

After a moment, Gilmore asked, "Any bad feelings, any second thoughts, Gray?"

Gray's answer was unequivocal. "None."

Then he added, "But I will plan properly next time. I only half thought through this one. Quick in and quick out has value, but there were too many details unsecured." His easy grin again lit the restaurant. "Do I get a decent debriefing?"

"No, you do not." Gilmore was also plain spoken. "A couple of those and your name will be a household word. Debrief yourself. You are trained to do it. If you have difficulty, particularly any tension, call me. Then we will do whatever is necessary."

Gilmore reported to the only other man who knew of Robert Grayson's employment.

"He did well, Chief. Recognized his mistakes and is ready to go again."

"No mental problems, no nail biting, Brad? I hate mental problems. If he shows the slightest tightening up, we take no chances; we dump him. You know that."

"I know, but not this time, Chief. Grayson is exactly right. He lacks some sort of governor that keeps the rest of us in line, making it hard for us to kill. He just doesn't mind. It is amazing, really, no passion involved, but genuine interest. He can dispatch a human as remorselessly as we could club a fish."

The Chief shuddered, "Scares the hell out of me."

Gilmore laughed, "It is not like that. Think of Grayson as a man who can read a book while walking a high wire. Otherwise, he is perfectly normal. He can do just this one exceptional thing that none of us can approach. He can use his talent as effortlessly as we can sit down.

"I like Gray. I like him personally, and I am going to be extremely careful how we use him, extremely careful, Chief."

7

Officer Bucky Larson sat on Gray's back porch, looking through the screen and across the mowed field. He had removed his pistol belt and propped his booted feet on a hard-topped hassock.

"Man-o-man, Mister Grayson, this is a beautiful place. A man could really relax here. If my ship comes in I'll have a farm just like this someday."

Gray chuckled, "This isn't much of a farm, Bucky. The crops just about pay the taxes, but I am glad you like it."

To his own mild surprise, Gray heard himself ask, "Would you like to see the inside? It's a little different."

"I surely would, Mister Grayson, there are all sorts of stories about this place."

"Oh, what do they say?"

"Well, mostly that this was a government bomb shelter of some kind, and you built a house over it."

Gray chuckled, "Close enough. You can decide for yourself. Two things, though. I would appreciate it if you did not describe what you see to just anyone. This home is special to me, and I would rather the public does not know my secrets."

"It's a deal, Mister Grayson."

"The second thing is that you call me Gray, I've never answered too well to Mister." Grayson smiled, "It makes me feel old."

"OK . . . Gray." Bucky worked through the familiarity. He buckled on his pistol belt as Grayson smiled again.

"God, Bucky, you modern cops carry more than Batman. You ought to have suspenders to support that equipment load."

Larson slapped his belt affectionately. You need it all these days. Pistol, two extra magazines, pepper spray, radio, handcuffs, usually your flashlight," His grin was engaging. "Looks menacing, doesn't it?" His radio squawked, and Larson adjusted the squelch.

Gray shook his head. "How can you run down a bad guy lugging that load?"

"Not too well, ah . . . Gray. The armored vest is worse, though. It weighs a ton, and in hot weather you hate it."

"Makes you look fat, too, Bucky."

Gray led the way inside, but Bucky stalled just within the entrance.

"My God, the walls are at least three feet thick, and the doors--incredible!"

"That's the way they build pillboxes, Buck."

Larson's brow furrowed, "You know, I've always wondered why they call bomb shelters pillboxes." His eyes were roaming taking in the expensive furnishings and manly comforts.

Gray gave him time by answering. "I suppose pillbox fitted because fortifications sat around looking like old time physician's rectangular boxes. Believe it or not, pill bottles are fairly new on the scene. Catchy name and pillbox caught on, I guess."

Bucky looked out windows and exclaimed at their size. He did not notice the rolled away steel shutters, and Gray did not point them out. Each shutter had a wooden facia that seemed part of the window frame, and the exposed rollers lay in an open trough that rarely caught the eye.

Larson sat in Gray's chair and marveled at the giant TV screen that appeared implanted in the wall.

"Now this is living, Gray."

"Yeah, but I have the same trouble you do, Bucky--there's nothing on the tube worth watching."

Gray did not take Larson upstairs or into the basement. They did examine Grayson's elaborate shop including the loading bench.

Bucky was again laudatory. "What a shop! You could build anything." He was specially interested in the handloading set up.

"Gee, you must have a hundred sets of dies, Gray."

Grayson sighed in resignation. "I've loaded about everything over the years. I sell off guns, but the dies stay here." Gray chose die boxes at random. ".38/56. I had a Winchester 1886 model in that caliber about thirty years ago.

Here's my old .218 Bee dies. I mentioned shooting that cartridge to you. The Bee was nice to load and shoot. Some cartridges are not as smooth loading. This .257 Weatherby is an example. I had to full length resize every loading, and it took forever to trickle the big powder load through the skinny neck."

Bucky understood the talk. He again vowed that someday he would have a real loading set up where he could experiment and not have to pack everything away when he quit for the day.

Larson noticed the lack of a bar. Most fancy houses he visited featured a wet bar with endless bottles to choose from.

Bucky hazarded his guess. "You don't drink, Gray? At least I don't see any signs of liquor around."

"No, I don't drink--never have. I don't socialize, so it is not necessary to have a lot of booze showing. But, I've got a bottle or two, if you want a drink." Bucky did not.

Larson wished he knew Gray's marital status. There was clearly no woman around. There were no flowers or tasteful knickknacks. The decor was lean and the kitchen showed little use. Gray could be divorced or widowed. The subject would surface eventually.

Grayson asked, "Did the restaurant get paid for what we were having when that punk stole my motorcycle?"

"I squared up the next day. Steve said yours was on the house for catching the stupid jerk."

Gray got them soft drinks, and Bucky listened to his dispatcher's chatter. He frowned in annoyance. "A domestic disturbance. I know the address. Hell, everyone of us knows it. The guy who weighs about one hundred and ten gets drunk and beats up his one hundred and eighty pound wife. She calls 911. We arrest his ass, but she changes her mind, and we let him go. Two days later, she gets drunk and stomps his bony butt. He calls us in utter outrage, and we do it all over again only the opposite way."

Gray said, "You need local police. That kind of stuff shouldn't be at state level."

Larson sighed, "Tell me about it. Those two have probably worn out a pair of handcuffs without one charge getting typed out. Lately we just lock up whoever seems drunkest.

Bucky queried, "You get into any of this during your federal service, Gray?"

"No, I was just a courier. I lugged stuff all over the world." Gray shrugged, "I never had apprehension power and never needed it."

"But you carried a gun back then?"

"Nearly all the time. Some destinations precluded carrying a firearm but not many. My pistol was for self-defense, not law enforcement."

"Ever need it?"

Gray laughed, "Well, I waved it in a few faces, and it made me feel a hell of a lot safer a lot of times."

Gray shifted the burden. "How about you, Buck? You had to shoot anyone?"

"Not yet, anyway, but I have come close more than once. It's not a good thought, but I am trained to do it."

Trained to do it. Grayson's mind closed around the words. Larson could not know, would not begin to suspect, how little he was trained for killing.

For a policeman, at any level, shooting someone was a rarity followed by mandatory time off, vast paperwork, debriefings, and psychological counseling. A cop enforced law, shooting anyone was improbable and a most minor part of his duties.

Grayson's had been the other side of the coin. Killing was what he had done, all else had been secondary, simply window dressing to disguise his real vocation.

Grayson's office floor manager had once taken umbrage at Gray's frequent absences and his failure to satisfactorily sign and transfer to his out box the requisite number of completed forms.

His bald head glowing with the heat of righteous wrath, his bulbous whiskey nose bright above quivering jowls, the manager shook papers in Grayson's face and threatened for all to hear that if Grayson did not get to work he would find himself on the street looking for another job.

The manager knew better. It was clear if unspoken that Robert Grayson was some sort of political appointee whose influence gave him a work station and a title without genuine duties. The situation was not unique. Within the vast Washington bureaucracy there were workers who had never been seen. It was the manager's duty to not notice Grayson's peculiar work hours and work load, but on this day, the floor manager believed himself overstressed and surrendered to a hunger to tell someone off.

Unfortunately, Robert Grayson had barely returned from his own stressful activity. He had no brotherhood to discuss with. He received no formal debriefing, there were not even any forms, and no psychiatric soothing and evaluating. The floor manager had picked the wrong moment. Grayson's fuse was also short.

In his adjoining cubicle, Dick began whistling tunelessly. The manager's office door slammed, and Gray stood up. Dick said, "Easy, Gray, easy now."

Grayson was soundless crossing the floor. He seemed to flow through the office door, and Dick recognized the training that closed the door without lock click.

The manager raised his head, his eyes red and angry. "What . . .?" He got no further.

Grayson's grip on his shirtfront jerked the manager erect and sprawled him across his desk amid a jumble of scattering paper and falling bric-a-brac. Blank astonishment gaped the manager's jaw as Grayson flipped him onto his back. Before he could gather his senses something hard and steel tasting was rammed painfully into his mouth loosening his upper plate and threatening to gag him.

Grayson's eyes, colder than ice flows pinned the manager, and their almost blank unconcern shuddered the man at least as thoroughly as the pistol jammed into his mouth.

The hard grip on his shirtfront shifted to his throat and squeezed off most of the suffering man's air. Grayson's voice was as merciless and as emotionless as his eyes, but the words bit with a paralyzing intensity that burned into the manager's guts.

"Do not ever raise your voice or threaten me again, or I will kill you as I would a fly. Believe me that your body would be immediately disposed of, and I would not even be asked about it."

The hand tightened, shutting off blood and breathing. The victim felt his consciousness fade. When it returned he was alone, still sprawled on his desk, his dental plate loose in his mouth, and his pulse pounding in horror and terror.

It was dream-like. It was inconceivable. He was . . . men did not . . . no one could . . .but it was real, and the manager knew with certainty that death had brushed him.

The taste of the oiled steel pistol was in his mouth, and he hurried to his private washroom. The mirror showed a face drained of color with staring eyes suddenly aware of the closeness of personal violence and death.

He rinsed his mouth and adjusted his clothing. Waiting for calm and order to return to his mind, the manager replaced his desk items, gaining solace from the simple organizing. His door boasted a peephole designed to identify those knocking, but Grayson's cubicle was angled away. The office appeared normal, but for the first time the floor manager appreciated the dangers that could lie within the almost unidentifiables occupying his cubicles.

The manager fingered his tender throat. Remarkably, his neck was unmarked. If he had rushed to complain there would have been no evidence.

The manager sat down. He would wait until Grayson had left the building. Then he would decide what to do about threats and intimidation including physical assault. He knew, of course, that he would do nothing. He would speak to no one. He knew absolutely that

Grayson would kill him, just as he had said he would. The manager shuddered, and his eyes teared. He tried to read the words on a paper he held, but they were meaningless. Now he knew what the devil was like, for as surely as he still breathed, he had met him face to face.

Dick said, "I've got an idea for a double wall system. Got time to look?"

"Sure." Grayson crossed to Dick's cubicle and pulled up the spare chair.

Dick's pencil flicked toward the manager's office. "You two reason together?"

Grays' voice held amusement. "Yeah, I asked him to be more courteous and restrained."

"Think he will come out soon?"

Gray looked speculative. "I think he might wait until we've all gone home. He's probably a little embarrassed by his outburst."

Dick's chuckle was deep with amusement.

8

The plywood sign pop-riveted to the trailer side said, Field Headquarters, County Militia, Colonel Buff Bolinger Commanding. The lettering was black stencil on a white background, *military*--as the Colonel saw it--a man who considered himself a real soldier unimpressed by fancy crap that did not advance the mission.

The County Militia had been activated following the federal government's fiasco at Waco, where civilians including women and children died. Buff Bolinger had almost single-handedly created his regiment. He had to have a regiment to rate a "Bird" Colonelcy.

Bolinger had advertised in the local papers for *"Genuine patriots, men with guts and guns."* He had held meetings where he determinedly proclaimed a federal conspiracy to sell out to the United Nations, to collect all privately owned firearms, and to imprison everyone who resisted domination.

The colonel had evidence, first in selectively chosen photography of a Waco video plus handouts describing UN marked track vehicles and unmarked black helicopters seen maneuvering. Foreign troops that were ready to pounce on an unprepared and unsuspecting United States populace were reported training in Idaho and Canada.

Bolinger diverted questions about where in hell the United Nations would get enough of an army to attack anything by pointing out that the whole world hated Americans and that the bastardly Chinese army had enough men to invade everybody.

An objector noted that five thousand or more friendly foreign troops were always attending American military schools and training with American units, but Bolinger was not listening.

At one meeting a retired soldier pointed out that the foreign and peculiarly marked tanks seen around were out of Gulfport, Mississippi and were part of a U. S. Army unit that acted as enemy on field maneuvers. But those eager to believe in government plots turned their attention back to Bolinger.

Most attendees departed early from Bolinger's meetings, but there were usually a few conspiracy buffs and a couple of thickheads who clearly saw the feds' secret plan. Bollinger took names and assembled his recruits for drills and meetings. By the death of Gray's motorcycle thief, the County Militia Regiment had swelled to fifty-seven members, more than half of whom could be depended upon to show up for assemblies.

Colonel Bolinger organized units as best he could. Because his numbers were small, he developed teams instead of squads. Because rank improved morale, Bolinger considered each team member a skilled cadreman available for further unit expansion--when the people wised up and recruits came pouring in. Master Sergeant was the lowest military rank Bolinger conferred, and there were clutches of captains and numerous majors.

Militia adjutant, Major Charlie Fobbs did the regimental paperwork. It was Fobbs who phoned every member the day before a militia meeting to get attendance up. It was Fobbs who had contacted the Northern Pittsburgh Militia who sent the motorcycle thief east as a liaison to coordinate the two militia's future field maneuvers.

Colonel Buff Bolinger had been both a Company Clerk and a Supply Clerk in an infantry outfit between wars. He did not mention that military record. Bolinger claimed to have turned down an opportunity to attend West Point and to have refused an offer of a field commission because there was no war to fight, and the army was not serious about what it was doing. Draftee Bolinger completed his two years mandatory service and got out.

For twenty years Buff Bolinger had run a small out-in-the-country gun shop. He espoused an ultra right wing militarism worthy of a Martin Borman, but where Borman would have flourished banners and slogans, Bolinger spoke of sinister plots, underground movements, and the need to organize before it was too late.

The idiocy at Waco insulted sensibilities throughout the nation, and for the first time Bolinger's carping found sympathetic ears. As a Militia Colonel, Buff Bolinger began his campaign to prepare America to meet the One World Conspiracy with guns and bullets. His regiment stood ready to act as the "hard core," the trained warriors, when "push finally came to shove."

Bolinger loved those kind of tough sounding clichés. He used a lot of them, and his few minions appreciated their Colonel's straight from the shoulder, good old American patriotic toughness.

Bolinger adored his colonelcy. He was seldom out of camouflaged field gear. He wore a government .45 pistol in a government holster with US stamped on the flap. He had a colonel's eagle above his automobile license plate, and he had laminated ID showing his unit, his rank, and his steely eyed photograph. It annoyed him that most clerks preferred to see his Pennsylvania driver's license.

The accidental death of the motorcycle thief embarrassed Bolinger. The death of Pittsburgh's representative was a stain on the County Militia's honor. Two of his men had been with the thief, and they should have provided protection.

Of course, the attempted theft had been dangerous, but the idea had been the thief's, and if it had succeeded, the rubbing of two highway patrolmen's noses in it would have been glorious. Cops were part of the hated and disdained government that was strangling the country.

Bolinger could when convenient ignore the detail that law enforcement officers spent a lot of money on his merchandise. Buff smiled in public, but vilified in private.

In a sense, Bolinger had instigated the attempt. He had mentioned at the outfit's monthly assembly that they should put together a trained motorcycle unit that could scout and run messages. He wondered aloud where they could acquire cycles. Their Pittsburgh visitor had seized the moment to grab the Harley, but the two louts with him had somehow let the civilian owner get through and caused the mess that followed.

Pittsburgh did not like it, and Colonel Bolinger could at least throw them a conciliatory bone.

For his conference with the two men who had failed, Bolinger opened the Regimental Headquarters. The trailer lay in a clearing within the almost one hundred acres of chopped out woodland Bolinger owned. Rusting concertina wire funneled visitors to the main gate which was usually padlocked against spying interlopers.

Colonel Bolinger had his two men standing at their best attentions, and he kept them there for longer than normal. The two needed to know that they were not yet forgiven for letting the liaison from Pittsburgh get killed.

Finally relenting, Bolinger ordered "At Ease" and the two relaxed.

Bolinger questioned, "So what did you find out?" There was hesitation so he added, "You first, Dillard."

Bags Dillard said, "Well, like you said, Buff, we . . ."

Bolinger halted him. "We're on duty now, Sergeant Dillard. Address me as colonel."

Bags tried again. "His name is Grayson, Colonel, Robert Grayson. He lives out beyond town, out along the river."

Fudge Crowder wanted in. "Guy is retired. Old buzzard, but kind of lean and wiry looking."

"He lives alone?"

"Yep, but people don't know a lot about him. Seems he was born here but moved away."

It was again Dillard's turn. "We didn't try looking, but he has some kind of a big house built over an old government lookout or fort or something. The place is way out all alone, Colonel." Bag's voice was suggestive.

Bolinger turned to Major Fobbs. "You getting this down, Charlie?"

"I've got it, Colonel. I'll learn more now that I've got a name."

Bolinger rose to pace. "You men are all square with the law?"

"We're square, Colonel. No sweat really. We just claimed we'd picked the guy up hitchhiking. What could they prove? We was out in an hour."

Bolinger nodded. "That was good, clear thinking." He assumed a good colonel's stance in front of them, feet slightly spread, firm-jawed, fists clenched on his hips. He wore his black beret square across his forehead in Willie G. style. He knew he looked a leader.

"Men, we've looked bad to our comrades in the Pittsburgh unit. We've let one of their men die. Majors Grove and Stuart attended the funeral as our representatives, but that does not square things. We've got to do more."

Bolinger moved behind his desk to grasp his leather covered swagger stick, the symbol of military command.

"So, Master Sergeant Dillard, Master Sergeant Crowder, it will be your duty to return our honor. It will be your duty to beat the snot out of the damned civilian that caused a good man to die.

"Now I know that you had to concentrate on delaying troopers, which you did, and could not guess the civilian's intent, but you men are involved. You were there, so it is fitting that you do what has to be done."

Crowder smirked, "Hell, Buff . . .er, Colonel, it'll be a pleasure. How bad do you want him hurt.?"

Bolinger was firm. "Do not kill him, for God's sake. Just smash him up, break his face, knock out some teeth."

Dillard repeated, "We can do that, can't we Fudge? Leave it to us, Colonel."

"How soon do you want it done, colonel?" Crowder intended to be prompt.

"Make it within the next few days. That will show Pittsburgh that we are a crack outfit that stands tall."

Bolinger watched their big wheeled pickup pull out. Good enough. His only worry would be that their enthusiasm might kill the old bastard. Not likely, and a chance that had to be taken.

Pounding out an old man could get Dillard and Crowder jail time but maybe not. They could claim, what was his name? Grayson had jumped them because he blamed them for the attempted stealing of his cycle. Bolinger did not really care, this would make him look strong to the

Pittsburgh officers. Hell, they had a General out there. Bolinger envied a rank like that.

Gray picked up the light-barred pickup when it swung in behind him on High Street. The two hulks silhouetted in the cab almost had to be the same gut tubs who had been involved in stealing his Harley. He turned downhill, then right on Main Street. The pickup was still there.

So, what did they want? If the two thugs wished to intimidate they would most likely ride his bumper, but they stayed well back. It was hard to believe that the two expected to follow unobserved--in that huge, one of a kind rig? But then, he was not dealing with high IQ professionals here.

Rough stuff? Grayson's mind groaned. How dumb were these guys? Real dumb, apparently. Gray's thoughts firmed. So be it. He experienced no other emotion. He never had.

Fudge said, "Hot damn, he's heading down the river. When he stops we'll get him."

"He might go into the city, Fudge."

"So what? Nobody in a city will try to help him."

"I know that, man. I'm talking about cops. We don't need no cops."

"Hell, cops are never around. Quit worrying, we'll get him wherever he goes."

"Well, don't follow so close. He could notice."

Fudge was disdainful. "This ain't some secret agent we're following, Bags. We're just another farmer's pickup to him."

Gray chose Elie's Steak house. At mid-afternoon the restaurant parking lot held only two cars. If the dumbos followed him in, only a few customers would have their meals interrupted.

Gray knew Elie's. The steaks were tender, there were real cloth napkins, and the steak knives were never dull.

He chose a table where he would not be pinned in; one far enough from other diners to feel separated. The waitress came quickly, and Gray ordered a sirloin, medium rare, new potatoes, cold slaw, and iced tea.

Through the window Gray could see his car, but the pickup was hidden. He unobtrusively shifted the .380 pistol to the front of his pants, tightened his belt a notch to make sure it stayed there. Stuffing a small hand gun behind a belt worked better in movies than in practice. They tended to drop down a pants leg during vigorous movement.

Bags grew restless. "Let's go in and get him."

"Relax, he'll come out. Then we'll lick him."

"Hell, Fudge, everybody passing sees us sitting here waiting. I've seen three people we know already."

"So what?"

"So we won't be able to claim the guy jumped us. Everybody will know we were waiting for him."

Bags was nervous. "I don't like the idea of busting him up right over here against the highway, anyway. Let's get him inside. We'll be gone before the 911 call goes in."

Fudge could see the logic. "OK, let's go get him. You going to sap him?"

"Nah, but I'm using my knucks." He pulled on a brassy looking set of metal knuckles.

"For Gods sake don't hit his skull with them things. You could kill him, and we ain't supposed to do that."

"I'll just work on his face. Give me three licks, and you won't have to touch him."

"Like hell, if we're doing this, I'm getting in my own shots. Just don't skull him."

Gray saw them approach the door. He wondered a little about what they would attempt but knew when the second man reached shoulder high to grasp the door edge. A set of brass knuckles glinted on his right hand. Gray continued eating as they entered and looked around. He even searched a little for special feelings because he had not often been the hunted. His pulse remained steady, and the steak still tasted good.

No one looked up as the two burly figures approached Gray's table. Gray ate with his fork, his left hand politely in his lap.

Looming over the seated figure, Bags said loudly enough to draw attention, "What did you call me?"

To bags astonishment Grayson looked up and spoke as softly as if in church.

"What I would have said is that you stink, you are stupid, and you are ugly."

The unexpectedness of the insults stunned and froze both attackers.

Fudge Crowder came to first. He snarled and snatched the front of Grayson's shirt. Half blocked by his helper's big body, Bags Dillard stepped aside seeking a clear swing. Instant rage burned through any caution that should have asked why his victim was so coldly unconcerned. All Bags Dillard saw was a face being held for him to hammer.

Grays' steak knife held edge up in his left hand sliced long and deeply across the inside of Fudge Crowder's wrist. Skin, tendons, veins, and arteries parted, and Crowder's grip went dead. Hauling hard, the sudden release staggered him backward, still unaware of his almost severed hand.

Released, Gray's body dropped, his head dodged, and Bags Dillard's slamming blow missed completely.

Without hesitation, Gray drove the steak knife into his attacker's thick torso. The razored blade went in to the hilt, and Gray heaved it upward.

The knife had gone in just under the attacker's thick belt. The handle jammed against a huge belt buckle, and sitting down with poor leverage, Grayson was unable to rip upward until the man's guts spilled onto the restaurant table.

It did not matter. Gray left the knife in the man's body and slid free of his chair. The attacker with the nearly severed wrist began screaming and waving his blood pumping arm around. The stabbed assailant studied the handle protruding from his abdomen, then knelt and rolled onto his side as agony took hold.

Gray shook the shock frozen waitress into awareness. "Call 911."

The cut-armed man also sank to the floor moaning and clutching his wound to his chest.

Gray watched for a moment, but the terror stricken fool seemed unable to help himself. Two customers threw money on their table and fled. Other customers simply sat.

Gray said, "Oh, hell." He twirled a cloth napkin into a strap, knelt on his victim's neck and knotted the band loosely above the slashed wrist. He twisted a spoon into the tourniquet and tightened until the spurting stopped. He placed the sobbing victim's good hand on the spoon and said, "Hold onto this."

A cook and the waitress cringed behind their cash register. Gray asked, "Is an ambulance coming?" He got nods for an answer.

"Do you have any blankets here?" It took a repetition to drag out words.

"Yes, in our apartment."

"Get two. Cover these men. It might help hold off shock."

The cook charged away, and after a silent moment the waitress asked, "What happened? I didn't see what happened."

Gray said, "You'll get all the story later. For now, tell the police only what you know. Don't guess at anything. Do you understand?" He again got a nod.

An ambulance wailed in the distance. Gray did not enjoy the sound. Standing around waiting for officialdom was not part of his experience.

He spoke to the two remaining customers as the cook returned with blankets.

"It will be all right now, but the police officers will want to speak with you." The man and woman nodded, fear and perhaps awe showing in their eyes.

Gray stepped into the men's room and returned his pistol to its holster. He washed his face and hands, but found no blood spots on his clothing.

When the ambulance and police cruisers wheeled in, Grayson was eating mints at the checkout counter.

9

Bucky Larson rang Gray's doorbell, but Grayson was slow in answering. Larson waited with some impatience. God, he always missed the good stuff. He had been directing traffic at the Farm Show buildings when the action went down. He had caught bits and pieces over his radio but had not known of Gray's involvement until the next day.

Local police had handled the matter. What had started the brawl was a bit unclear. The cut and stabbed claimed Grayson had shouted an insult then used his knife. The set of brass knuckles on Bags Dillard's right hand cast serious doubt on that story.

Witnesses were confused, but they mostly agreed with Robert Grayson's version that an attacker out of nowhere accused him of saying something, jerked him by his shirt and tried to hold him while the second swung brass knuckles at his head.

Grayson had then defended himself with his steak knife. And, yes, Grayson did recognize the two as bystanders during an earlier attempt to steal his motorcycle. Mister Grayson had heard their names but did not recall them.

Gray came to the door wearing his customary slight smile.

"Hey, Bucky, come in. Sorry I was slow. I was at the handloading bench."

Grayson pointed the way but waited to lock the door behind the trooper.

He again led saying only, "I'm being careful. Dillard and Crowder might have friends."

Bucky refused a soft drink, and Gray went back to his seat at the loading bench.

Larson's intent to pursue the knifing was delayed by Gray's work.

Bucky gazed in awe at a huge cartridge Gray offered for examination.

"My God, what is it? What a monster."

Gray chuckled appreciatively. "I am one of the few in the world who load these. Cute little devil, isn't it?"

"I guess it's a .50 caliber of some kind."

"Yep, and a special one." Gray was obviously prepared to talk about it.

"Back in the 1950s, the army had this cartridge made up to match a recoilless rifle's trajectory. The gunner pumped out a few of these cheap .50s until he got on target, then he touched off the big gun and its round went to exactly the same spot.

"Later on the weapons system was dropped, and I bought a few thousand surplus rounds. Then, I had this built." Grayson opened a wooden rifle box laid out on a work bench.

Again Bucky muttered, "My God."

Gray handed him the rifle. "It's heavy, Buck. Weighs twenty-one pounds. The action was made from scratch, but it is basically an over sized Model 700 Remington with Pre-64 Winchester extractor and safety. I tried to get the best of both types." Gray extended the rifle's bipod for support, and Bucky sighted through a window.

Gray answered before Larson could ask. The scope is also one of a kind. It is a variable power 4X to 12X. That is not unusual, but the Mildot reticle is good out to 1500 meters and the elevation adjustments also work out that far."

Grayson was apologetic. "See this extra knob and turret? That's a bullet drop compensator. Once you estimate your range it allows fast adjustment, but it isn't much good right now. I keep changing my ammunition, so the trajectories never match the compensator."

Larson was nearly mesmerized. They discussed recoil, and Gray described the built-in reducers and absorbers. Bullets and powder took more time. Bucky's radio did not work too well inside the shop, but Larson ignored it.

Eventually, they again sat on Grayson's porch. This time on the front porch with the trees crowding in on them.

Gray explained, "I looked up, and there they were. Man, that bigger guy, Dillard, I guess, swung at me with those knucks so hard he about fell over. I just stuck the knife straight into him."

Larson's chuckle was joyful. "You know, Gray, we investigate a thousand cases of assault, and almost never does the innocent guy win. Mean guys often pick victims they can be sure of. For once, it came out right."

Larson was quizzical. "How come you didn't shoot them, Gray? You had your pistol, didn't you?"

"Hell, Bucky, they were on me while I was eating."

"Then, according to the report, you put a tourniquet on Fudge's arm. You probably saved his life."

"I probably did. He was just sitting there letting himself bleed to death. Nobody else was moving, so I did it."

"And got blankets put on them both." Larson's voice turned inquisitive. "Are you used to this sort of thing, Gray? I mean have you been around a lot of death?"

"I was an infantry rifleman back in the Korean War, Buck. I was only a kid then, but I shot straight and didn't get rattled. Likely some of that ability stays on."

"You've had training since then, haven't you, Gray? I mean the way you slipped up on me when I was hunting, and you've got a balance that I've seen in good athletes."

"And you've got a good eye, Bucky. Yeah, I've been trained, all kinds of schools and skills. It seemed as if when I wasn't delivering something I was off to some hard-assed school learning something I hoped never to use."

Bucky was respectful of education. "How high did you get in this courier thing, Gray? I mean is there rank or something?"

"I know what you mean, Buck. I retired about the same as a GS-14 or a Lieutenant Colonel. I was as far as I could go, and I was sick of flying around with an attaché case chained to my wrist.

"A lot of my work was mindless. The government should have assigned Military Police Corporals to most deliveries, but it added status to have a ranking courier delivering your stuff."

Bucky wanted stories, but Grayson could not even hint at the real ones. He could not safely lie about being able to handle himself. Larson would sooner or later recognize his abilities. It was safer to continue with the same old courier cover that had withstood investigation throughout his career--except for a time or two.

+++

They sat in Brad Gilmore's car studying the building a block away. It was a five story concrete structure with only front and rear entrances. No windows appeared below the top floor

"He's there, hiding out in the penthouse, Gray, and we haven't a chance in hell of getting at him."

"Storm the building?"

"In the middle of the city? Anyway, the place is a fortress. Only two elevators going up, no stairways at all. Armored doors everywhere, at least fifteen highly trained security men with big weaponry, and possibly pre-set explosive charges all over the place.

"The roof?" Gray thought it best to eliminate even the most obvious.

"Concrete and steel with emergency firing slits that could rake the top clean. There isn't even an air duct to hide behind up there."

Gray chuckled, "Maybe a B-52 bomber could accidentally unload on it.'

Gilmore remained serious. "We can't create an obvious international incident, Gray, and sitting on top of a hill like that building is you couldn't get in a shot with that .50 caliber gun you've got, unless he came out onto the terrace, and no one has ever so much as opened the door up there."

"You're sure he is inside?"

"Positive, and they know we know. Right now they'll be questioning him politely, and as long as he is cooperative they'll stay polite, but once he has told all of the things he thinks he knows the hard stuff will begin, and Jules Chalmers will discover the huge mistake he has made."

"What can he tell, Brad?"

"Everything, Gray. The man knows names and places. We figure we've got about three hours before we have to shut down teams and operations all over the Middle East. We doubt the Iranians even guess how much Chalmers knows. That's why we have a little time.

"The bastards have an antenna they stick out a window, but we know their code, So far, they are just pleased as punch over Chalmers walking in off the street, but they will get down to serious business pretty soon."

"Man, I hate these hurry-up jobs. How am I supposed to do what the entire agency believes is impossible?"

Gilmore shrugged. "They don't know about you. This is all my idea. As far as I know you are the best at this kind of thing. I thought maybe--well, you know, just maybe."

Gray realized suddenly that there was a maybe. If he could . . .

"Do we have a layout of that fifth floor, Brad?"

"Not a real one, just a calculation based on window placement and the known elevator shafts. There are no plans, and no one we have has been past the ground foyer."

"Hell, somebody built it."

"Right, and we've picked all the brains we can find, but the plans are gone, and even the cleaning staff is Iranian live-ins. They come out in groups and never separate. Luckily their women dress western. Those black hoods would hide everything, and we wouldn't even know who was who."

Gilmore sounded hopeful. "You got an idea, Gray?"

To give himself more time, Gray asked, "Have you considered pumping about a dozen tank artillery rounds straight in the windows? Cooped up in a concrete box like they are, inside explosions would do in everybody."

"You're nuts, Gray, that would be worse than an accidental bombing. Whatever happens cannot be an obvious United States Government action. God, half the world is waiting for us to commit some terrorism to justify their own rotten activities."

Gilmore sneered, "Believe it or not, one idiot did seriously suggest working a cruiser up the Potomac and driving a five inch shell into the top floor."

"God."

"Yeah, we could tell the Washington Post that we just happened to have a cruiser north of the Woodrow Wilson Bridge which just happened to shoot a five inch gun whose shell accidentally took the lid off an Iranian diplomatic building. And Gray, it would only take a few days to get a cruiser into position."

"Sounds reasonable to me." Gray chuckled, "Did they promote the guy or fire him?"

"Probably promoted the idiot. If President Carter heard about the suggestion he would either die of a heart attack or invite the fool to the White House for serious consultations. It'd be a toss up."

Gray said, "All right, Brad, I've got a plan. You're not going to like it, but it's my only offering."

Gilmore did not like it. He said, "I don't see how . . ."

"Well, it'll work or it won't."

"God, Gray, that's trusting to a lot of luck. I mean a lot of luck."

Gilmore studied his protégé. "I'd like to declare it too dangerous, but we are desperate."

"OK, Brad, here's the deal. It'll be dark in half an hour or so. You make sure that one street light doesn't go on. Just that one on the corner, Brad. More could arouse suspicion."

"I don't know if . . ."

"Next, put a Sniper in whatever stakeout you are using now. Don't hunt for a new position. All he will do is make sure no one pots me while I am leaving the scene. It could get close right then."

Gilmore groaned, "We're not rigged for this fast stuff."

"Neither am I, but you said we have no time."

"What will you be doing?"

"Getting ready, and I'll start one hour from now. Have your car right here so I can check in before I begin. I'll need to know if there's a change."

"God, Gray, I don't even have the right to approve what you are proposing, and I'll never get authority in one hour."

"Your choice, Brad." Gray's grin was a little ghastly. "If I make it, you are a hero. If I don't, just claim you couldn't stop me."

One hour, both men went to work.

Gray tapped on Gilmore's window, and the handler opened the passenger door.

Gray said, "I can't sit down with this rig on. Is the Sniper ready?"

Gilmore squirmed. "Well, they got a mini-14 to one of the stakeouts. He says he can shoot."

Grayson groaned aloud.

"Damn it, Gray, we don't have Snipers standing by. Neither does the FBI, and just hinting that we might be trying something on their turf would really turn on their alarms."

Grayson said, "To hell with it. I'll be back in a blink, anyway. Make the signal."

Gilmore turned on his headlights, blinked once, and shut off. Up the street a heavy car left its parking place, veered wildly, and crashed into another parked vehicle. Both drivers were instantly out, cursing and shoving. Gilmore said, "Go," but Gray was already at the building corner.

The diplomatic structure's exterior was stuccoed concrete, but each corner had been enhanced by imitation concrete quoin stones that projected four inches from each wall surface. Each artificial stone was two feet high and separated above and below by nearly a foot of normal wall. The extended corners were to be Grayson's ladder to the fifth floor.

Gray appeared from behind decorative shrubbery planted along the sidewalk. He rose almost wraith-like spidered against the corner, each upward step stretching his reach and forcing his body away from the wall's safety.

The climb would be tricky under the best of conditions. A shorter man could not have reached the upper lip of the next stone, and the almost three foot step up resulted in a spraddled and awkward grip before Gray could release a hand and snatch another hold before his weight threw him backward.

Brad Gilmore found he had forgotten to breathe. Grayson was a darker form obvious to anyone chancing to look up, but each step improved the angle, and if he could remain unseen for the first dozen steps upward, Gray would be above casual notice.

The fight in the street had turned physical. Two combatants were on the ground flailing, and a third party was apparently kicking both. Damned convincing, Gilmore decided. He recognized no one, but they were his people.

Gilmore could make out a rope coil behind Gray's belt, but his killer had no obvious pack of equipment. How . . . ? Grayson had not given much detail.

It was harder than Gray had expected. The quoin stones had been shaped with a downward sloping top edge to run off water. His only grip was at a stone's corner, and his footing on the downward slant was slick and forced his balance away from the wall. Committed, he had no comfortable choice, but it was reasonable to suppose that if he could handle two steps he could manage two dozen.

Grayson worked at it. He forced concentration. His mind did not consider his exposed back or the deadly fall. His pulse rose from effort, not from debilitating fear. His body heated, and the thought of sweating hands could have been defeating, but Grayson's talent held. His mind stayed focused, and each step brought him closer.

There it was, almost suddenly the small railed balcony came into view. Still Gray worked higher.

When he could look straight across to the balcony floor, Gray paused, pressing himself tightly to the corner. Balancing carefully he tentatively released a hand but immediately felt his weight begin to shift backward. Gray regained his grip. He would have to be fast, smooth, and successful the first time.

Brad Gilmore feared he could not stand it. Grayson clung like a fly almost to the flat and featureless roof. Because of a wide overhang, Gray could not go much higher. Well beyond reach to Gray's left lay the relative safety of the balcony. Gilmore's knees rose almost to his chest in tension.

Gray's rope was hung to his belt by a five clawed grappling hook. Twelve feet along the rope Gray had knotted a large hand loop. His task was to grip the loop and toss the grapple over the balcony railing. A simple toss on the ground but a hellish challenge while slowly falling backward. For the first time, Gray felt tension sweat pop. His clinging hands immediately dampened, and he had to move.

Gray pulled himself tight and instantly released his left hand. He concentrated on gripping both loop and grappling hook. He got the rope clear, only distantly aware of his balance going bad.

His mind saw only the balcony's iron railing. He did not wonder if the rail would hold or if his rope might snarl making the grapple fall short. Concentrating mightily, Gray made his toss, saw the grapple fly and felt himself toppling backward into oblivion.

From below, Brad Gilmore could not see the hook's flight, but he saw Gray fall, backward, feet still against the wall, then torn free, and . . . swing into a one-armed dangling arc below the fifth floor balcony.

By God, by all the gods, Grayson had done it! Grayson, by God, could do anything! For the first time Gilmore had genuine hope that Gray might succeed.

Gilmore saw the wild swinging halt as his man got a second hand on the thin line. Using his feet against the building wall, Gray went up in powerful hand over hand lunges. He reached the balcony and was instantly over and out of sight from below.

On the small balcony, Gray crouched low. His lungs heaved and his fingers were slick, but the strain was physical exertion not fear or unfounded relief. If asked, he would have recognized pleasure that his cast had been true and that the plastic coated grapple had held. His fingers stung with the burn of gripping the three eighths nylon line. Thin line was a horror to hang onto, but the climb using his feet aginst the wall had been short, and the burn would subside.

Gray listened at the curtained balcony door and heard English being spoken just beyond. The voices were in conversation with only an occasional word loud enough for recognition. Perfect, he was undetected, and it was likely his target was one of the speakers. Why else would English be used?

Gray reeled in his rope and removed the tied-in loop. He knotted the one hundred foot line's free end low around a baluster and placed the coil ready on the balcony floor. The grapple would provide weight when he went down.

Gray studied the glass door. Nothing special, it was an ordinary contractor's grade sliding patio door. Locked, of course. He did not even try to open it.

Gray checked his weaponry. He had little, but even that had hindered his climb. A 9mm semi-auto pistol was snugged to his back. Each side pocket held a military baseball style fragmentation grenade, and stuck into his shirt was his only entry equipment. Not much, but all he desired was one decent shot at Jules Chalmers.

Gilmore's photo of Chalmers had shown a jowly face with a Stalin mustache. Chalmer's had receding reddish hair and almost invariably wore a gold choker chain at his neck. There should not be more than one in the building who would look like that.

Gray withdrew his shirt packet and went to work. He had prima cord, putty, and a fuse device. Gray stuck the explosive cord near the glass edges all around the door and made a diagonal cross of what remained. He inserted his detonators, pulled the short firing wire, and pressed face first against the concrete wall as far away as the balcony allowed.

When he pulled, Gray recognized the tiny pop of the detonator activating, and he began counting. Like hand grenades, there would be a long four second delay. He drew his pistol and held it close to his chest.

Gilmore saw Grayson's black silhouette working against the light of the curtained glass door. He saw Gray crouch to one side, but he was still caught by surprise when the door exploded in a blast of light and . . . silence. The sound of the explosion, at least in Gilmore's direction, failed to carry. The handler barely saw Grayson move.

The rule was that when you heard the sound you were already late. Gray knew the drill; any bits or chunks coming his way were ahead of the explosion's noise and already past.

Gray hurled himself through the gaping hole where the glass had been. No shards remained to spear him, but he leaped into a mostly darkened room with lamps blown out and only half light to help him.

A figure rolled away, and Grayson shot it twice. A woman stared straight at him, and Grayson swung his muzzle to two men frozen rigid on a long sofa. Both had mustaches, but only one could be Chalmers.

Gray's first bullet struck Chalmers in the head, and Gray swung to the Iranian who held up a protecting palm. Grayson's shot went through the hand and into the man's chest, and Gray fired two more into the man. There was shouting and figures in the hall. Too many, Gray emptied his pistol into them and reached for both grenades.

The woman! She had gotten her purse and something was coming out. Her eyes glared fiercely into Grayson's, and he sledged a grenade into her skull.

A hail of gunfire roared from the hall, but no one charged. Gray pulled the pins on both grenades. He let the handles fly away, stuffed one into the possibly already dead Chalmer's vest, and hurled the second into the hallway.

Gray went out the door and onto the balcony in a surge. He tossed his rope and was stepping over the rail when the balcony entrance erupted in flame and smoke.

A bullet from somewhere chipped stucco into his face, and even in his hurry Gray knew its source.

Gray hooked the line over and under his clamped feet and started down. A second bullet struck just above him, but no more followed. He descended too fast, almost in free fall, unable and not daring to slow. Friction burned through the palms of his hands, and he could feel it wearing through his shoe protection, but he kept going.

For a lifetime, stuccoed concrete swept past, then he hit hard and attempted his best parachute landing fall. He was behind bushes, but he had landed badly, and an ankle both burned and numbed. Gray gained a crouch and managed a limping run. Gilmore had a car door open and the engine running. Gray fell inside and slammed the door. Gilmore wheeled away and turned immediately into a side street. They did not speak for a block.

Gilmore started, but Gray cut him off. His voice was as cold as the handler had ever heard it.

"That God damned Sniper shot at me at least twice."

"What?" Gilmore was clearly thunderstruck. "Why would he . . ."

'Take me to the stakeout right now, Brad. I'm going to find out."

Gilmore's voice approached panic. "No, no, Gray, give it a minute. I'll find out about the Sniper. Look, you are away clean. No one on our side can identify you. There are at least four men in that stakeout. If they see you, one or more will sooner or later find out who you are. You'd be done then, Gray, and possibly in danger."

The handler was right, Gray knew it. Others would have to do his work.

"All right, Brad, but you locate a phone right now and find out. I've never appreciated friendly fire coming in on me."

Gilmore watched for a telephone booth. "Did you get him, Gray? They'll want to know."

"Unless there was a look alike, I got Chalmers. One shot in the head and a grenade against his chest."

"God."

"There were other people, Brad. One was some kind of a fighter. He was evading when I shot him. A Caucasian woman, I bashed her head in with a grenade, and she probably got a load of Chalmer's blast."

"God."

There was another male, well dressed, mustache. I shot him, and he surely took some grenade shrapnel."

Gilmore's voice was weak. "Anybody else?"

"Probably, there was a bunch in a hallway. I emptied my pistol at them then threw my second grenade down the hall. It must have done some damage."

Gilmore saw a telephone and pulled up beside it.

Gray heard Gilmore say, "Successful, but this is a public telephone." Gray could not hear anymore.

When they started off, Gray said, "My ankle is swelling. I'd better get after it."

Gilmore was solicitous. "We'll drive straight out to Waldorf. No one could make a connection from there."

"Brad, that's a half hour drive. My ankle will look like a balloon. Hell, it might be broken, I landed with it twisted."

"We can't take a chance by seeing a doctor here. Everybody in Washington will be looking for whoever climbed down that rope. God, Gray, a hundred people looked up when that last blast let go. I couldn't even hear the entry explosion, but those grenades got attention."

"What about that Sniper?"

"That damned Sniper was just an agent who did not understand his orders. It was pointed out to me that if he wanted to shoot you he could have done it easily on your way in."

Gray snorted, "Give me a break, Brad. Before I did the job?"

Gilmore sighed, "Look, Gray, the senior man, who says he personally can't shoot a rifle that well, leaned over and clubbed the shooter. May have fractured his skull, in fact. Now those two have been together on other assignments. The senior says the junior just didn't understand. With that semi-automatic rifle working, the senior agent did not even have time to talk. He just leaned over and smashed the shooter with his spotting scope."

It was Grayson's turn to sigh. "It can happen, I suppose, but getting shot at on that rope felt real bad."

"How are your hands."

"Raw as hell."

"Don't mention them at the clinic."

"I know, I know."

There was silence for a while.

"You were a real James Bond back there, Gray."

"Yeah, I was great. Do I get a promotion?"

Gilmore was thoughtful. "You know, Gray, until tonight I have never seen an operation go down. I have ordered many, but they were distant, almost academic to me. Tonight? Tonight I was scared stupid. Even if I had the skills, I could never have attempted what you did."

Gray said, "You told me it was my talent, Brad. You've got yours, I've got mine. I am able to remain cool under fire, that's all." Gray chuckled, "You didn't mention my promotion."

Gilmore laughed, "Well, here's what we will do. First, you are leaving the country and your special duty. I do not want anyone bandying the name Grayson about. Where do you want to go?"

"For how long?"

"At least a year and perhaps two."

"Make it Saint Thomas."

"The island, in the Caribbean?"

"That's the place."

"What on earth can we have you doing in a place like that? I had in mind Berlin or Tokyo."

"Huh-uh, Saint Thomas. I've earned it. Get me one of those government grants to study the effect of multihull sailboats on the developing charter boat industry.

"What in hell is a multihull?"

"It's a catamaran or a trimaran, Brad, two or three hulled boats, and I know just the one I want."

"You want? You're not getting a boat. How would I . . ."

"Make it a Sea Runner 37, Brad. Ask around, someone will know."

"Damn it , Gray."

"We'll discuss promotion later. Better than James Bond, Brad, is that what you said?"

10

They rested on Gray's wicker chaise lounges studying the field beyond their toes. Bucky had arrived during Gray's morning stretching and had decided to join in.

Gray said, "You can't exercise with me anymore, Buck. You make me work too hard."

Bucky swallowed Mister Pibb, "I guess you are too tired to sweat cause I'm doing it for both of us."

Gray said, "Heaving your bodies around is harder for big men. You guys lift a lot of extra bone and muscle."

Bucky pondered, "let's see now, us big guys can't run as well as skinny people, we can't heave ourselves around, and we don't fit through small openings. What are we good for, Gray?"

"Oh, lots of things, Officer Larson. Big guys can hold things while we decide where they should go, and large men are terrific for delaying enemies while we get away."

Bucky snorted and finished his Pibb. "I'm not sitting here just to get insulted. Let's go get your Harley." He spoke while rising. "You sure it's ready? I hate going all that way for nothing."

"It's ready. Jim Howell called to tell me."

"A bike is never right after it's been wrecked."

It was Gray's turn to snort. "Who in hell told you that? Some of the finest custom Harleys are made from wrecks. Hey, I haven't told you all that Howell did to the bike."

"Like what?"

"Well, we got a new fork and all, of course, but then we got serious. Jim took the Harley entirely apart. I got a new Pro-1 frame that can take a seven inch tire and still keep the belt drive. We put on new laser cut wheels, extended tanks, and that new Corbin seat with the adjustable backrest. You wouldn't know the bike. It's lowered, and if a part could be chromed it was."

Driving out Gray's long dirt road, Bucky parroted, "Chrome don't get you home."

"No, but a Mikuni carburetor and two into one Bad Dog pipes might."

"Hey, Gray, you sure you brought enough money for all of this?"

"If I come up short I know you'll be there, Bucky."

Larson grinned, then said, "By the way, Bags Dillard isn't healing very well."

Grayson was clearly not concerned. "He's lucky to be alive." Bucky could not know how sincere Grayson was.

Gray said, "I am surprised they came after me, Buck. Hell, I never even spoke to them. They must have been closer to that cycle thief than they admitted.

Bucky nodded. "Had to be, but the only connection the investigation uncovered was that all three are militia members. The dead guy out by Pittsburgh and Dillard and Crowder in our local County Militia."

"What about this militia, Buck? Until now I'd never heard of them around here."

"Well, they don't amount to anything, Gray. I doubt we have anyone inside their outfit, but we probably know about everybody that belongs. I suppose you would have to label them right wing because their literature says all the usual things about patriotism, the constitution, Randy Weaver, and Waco.

"There's a lot of hate stuff in their paper and their reading list is so radical any sensible man would puke. They are a paranoid pack of losers and malcontents that talk with each other and listen to wild foolishness on short wave radio until that is all they know."

Gray nodded, "I know them. Some places they are racists, others believe the United Nations is training to take over our country. A lot of wannabe soldiers sign on as well as a few dumbheads who would join anything just to be included."

Bucky laughed, "That's them, and Dillard and Crowder fit about all of those categories."

Larson slowed his truck. "In fact, just up that dirt road on the side of the mountain is the County Militia's Field HQ. They've got a concertina wired compound with a house trailer and a flagpole.

"Hell, Gray, there can't be more than a couple of dozen of them. They put on a demonstration one time. Ludicrous! Out of shape people playing soldier with a weird assortment of AK47 clones and hunting rifles."

"Who runs the thing? Every offbeat organization has one character who holds it together, and that includes churches and social clubs."

"You've got it. Our militia has a self-appointed colonel. Off duty, Buff Bolinger owns a gun shop. A lot of men with similar interests are into guns. Militias pick up recruits at gun shows, or try to anyway. Most citizens just ignore them."

"Well, I'm ignoring them. If Dillard and Crowder are typical I'm not joining."

Larson chuckled, "Hey, Gray, if they knew you had that big .50 rifle Colonel Bolinger would be recruiting you himself."

Bucky turned the conversation. "Where do you shoot that cannon, Gray? I'd like to go along some day."

"I've got an in down at Quantico. They shoot a lot of Barrett .50 calibers down there, and my rifle interests them. If I go down, I'll give you a call."

Larson said, "I'd like to see your gun collection sometime, Gray."

Grayson did not answer. Yeah, Bucky would like to see, and his eyes would pop.

When Gray had decided to call it quits he began assembling all the weaponry that he figured he could need. Some he requisitioned and wrote off as used or destroyed. Much was illegal to possess without federal permit. More than a little had already accumulated and had been already stored in his cabin along the river.

Gray's armory was in the basement. A dehumidifier kept the room dust dry. Entry was by a bank vault door purchased used, and Gray's vault walls were equal in strength and thickness to any bank's.

The armory was well lit and conventional new and old rifles were racked and lying about. Those Bucky could see. Work benches and tools were handy. Heavy shelving was deep in boxed ammunition. Any gun man would be interested.

The real weapons trove lay hidden behind a built-in rifle cabinet. If you knew how, the entire unit could be swung like a door exposing a smaller secret room that bristled with armaments that could imprison a possessor for ten years plus a huge fine.

Gray had automatic weapons of many brands. There were cases of military grenades, flares, and signals. A half dozen LAWS rockets were hung on pegs, and a 3.5" Rocket Launcher was placed atop cases of its ammo.

Short-barreled machine pistols lay beside sawed-off shotguns so shortened they rivaled the pistols. In one corner a flame thrower waited, its tanks still painted military OD. Cans of fuel thickener and igniting matches were nearby.

C3 and C4 explosives were hermetically sealed, and a dozen forms of detonators were foam packed but ready. Blasting cord, called prima cord was available in rolls, as were coils of commercial time fuse.

Grayson had no known need for the munitions, but he had spent his life among such tools. He had been active for thirty years, and in service for a few more. Good items accumulated, and when he decided that it would be necessary to feather his nest he had included armaments otherwise impractical to obtain.

Gray had even kept his company pistol. He had never been officially issued the .357 Smith and Wesson revolver, but Brad Gilmore had gotten the powerful pistol for him back in the 1950s.

When his new handler had removed Gray from the field, he had demanded the pistol's return.

Gray had asked, "Do you have my signature for a pistol?"

The handler who genuinely hated Robert Grayson and the other dinosaurish baggage still part of the *new* agency, said, "Of course not."

Gray asked, "Do you have a purchase order for a pistol with this serial number, or is it listed as an agency piece?"

"Of course, not, Grayson. We never . . ."

Gray cut him off. "Then you do not get the gun Waverly."

They met in the dark at a rest stop well out on Route 66. Gray had chosen a picnic table out of direct light, and Waverly approved. He despised meeting with Grayson and was pleased that this would be his last time.

Grayson sat easily on the picnic table's end with the angry handler facing him. Waverly was a stout man, padded with too much flesh. In the dark, Gray could not clearly see the man's features, but he knew Waverly's head would be turning beet red with rage and frustration.

Gray had enraged Waverly with some regularity. Gray believed the handler to be a self-righteous ass unfit for his tasks and a danger to his operatives. Gray paid little attention to the man's wishes or demands. Now that he was being permanently removed from the field, Gray guessed it was time he illuminated a few pertinent facts for the pompous fool.

Waverly's voice jumped an octave and became a womanish shriek.

"You listen to me, Grayson. You . . ."

Grayson did not listen. What Waverly needed was a bit of the treatment Gray had given the recalcitrant floor manager years before.

Waverly felt himself grabbed, tipped off balance, spun, and slammed back onto the picnic table Grayson had been sitting on. Something hard jabbed him painfully beneath the chin, and a fist tugged on his hair. Leaning close above him, Grayson spoke, his voice as normal as usual. But Waverly heard more than the words. He registered the clicks of a cocking revolver hammer and realized with horror that the object grinding his chin was Grayson' pistol muzzle.

As if discussing the weather, Gray said, "Of course, you told no one you were meeting me, Waverly, so I can kill you in perfect safety." Then Gray added, "If you irritate me any more, I will kill you. I want you to believe that because I genuinely mean it."

Waverly, one hundred percent genuinely believed him. If old Gilmore's claims had been true, Robert Grayson had killed with perfect consistency for thirty years without ever failing or showing the slightest hint of concern about it. Oh yes, Waverly believed.

The pressure of the revolver left Waverly's chin, and he was raised into sitting by Grayson's grip on his hair. The night was chill, and Waverly was appalled that he was drenched in a sweat rapidly turning cold and dank.

Grayson said, "After tonight, I never expect to see you again. Do not interfere, and I will continue couriering until I decide to quit."

Gray again paused, then released the handler's hair. "People like me are all alone, Waverly. We have to take care of ourselves because people like you have no loyalty and would, without qualm, betray us to anyone handy.

"That used to be an occupational hazard, but as I grew older I prepared for a moment just like this. A moment when I could not only no longer trust my contact but would find it necessary to tell him so. You could say, Waverly, that I have been preparing ever since you came on the job.

"Now, what I have accomplished, Waverly, is to chronicle in exacting detail all that you have asked me to do and all that I have done.

"Do I sense doubt?" Grayson's hand came into view. It was held close beneath Waverly's nose, and with utter surrender, Waverly saw the compact tape recorder.

Grayson said, "I have made copies, of course, and to while away long hours at my desk I have transcribed everything. Finally, just before I came here I activated a plan which mailed copies to many places. As long as I am well and check in regularly, no packages will be opened, but

Waverly, if those packages are opened, killers from a dozen nations will have your name and be hungry to rip out your guts."

Gray stepped away, but Waverly feared to move. After lengthy silence Grayson said, "Killing has never affected me, Waverly. I have never understood why, but it does not. If I decide it is the right thing to do, I need not hesitate for fear of second thoughts. It might interest you to know that I have killed a few deserving that you did not order." His genuinely amused chuckle scalded his listener's nerves. "I am a poor choice for mugging, Waverly."

Then Grayson was gone, and after a moment, Waverly heard his car start and move away. Only then did the handler rouse himself. He swiped away cold sweat and straightened his clothing. James N. Waverly planned to walk away. He intended to never see or mention Robert Grayson again, and except for once, long after, he never did.

+++

Bucky said, "I'm going to bring my mother over to meet you."

"What?"

"Hey, Gray, she is one hell of a nice lady. You need someone to squire around on that Harley. Mom would love that, and you had better watch out, she is a dead shot with a pistol."

Grayson's voice was weak. "A dead shot?"

"Yeah, my grandfather taught her, and he was something else with a hand gun."

Gray got himself together. "Now hold it, Bucky, I don't need any fixing up. I can get my own dates. Anyway, your mother will not be thrilled to hear that her son is setting her up with some old river hermit."

Bucky was not dissuaded. "My dad's been dead for five years, Gray. My mom's got too much on the ball for anyone available around here." He added with a young man's brash enthusiasm, "You're both getting on in years, and you need company. I'll bring her over one of these evenings. I'll tell her about the Harley, and I've already told her about you."

Grayson said, "Oh, for God's sake, Bucky." But he liked the young cop, and he was pleased that he was liked in return.

Bucky said, "Her name is Anne, Gray. Just so you'll know."

11

Brad Gilmore had asked Gray to meet him in a Baltimore restaurant noted for its seafood. When Gray arrived, Gilmore was elbow deep in a pile of cooked crabs beating mightily with a wooden hammer. At Gray's approach a man and woman excused themselves and walked together beyond earshot. Grayson was surprised because Brad Gilmore had never included anyone in their meetings.

Gilmore attempted to clean a hand for shaking but gave it up and waved Grayson to a seat.

Gilmore saw Gray's eyes follow the couple. "It won't matter, Gray. Those are my daughter and her husband. Good cover for a last meet."

"Last meet?" Grayson was genuinely startled.

"Yep, I'm out. Being forced out, actually. Too old, know too much, can't handle a computer. The usual reasons for changing the guard."

"Come on, Brad, you aren't that old." Gray's words were automatic, but when he thought about it, Gilmore was old. Hell, Gray was fifty-two. Brad had to be at least a dozen years older.

Gilmore's head shook. "No, I am old, and they're right. My God, Gray, I am damned near a charter member of the agency. The only reasons I've been kept around are that I don't do much, and there is real reluctance to assign new men to my kind of work." Gilmore's glance was sharp. "Or your work, either, Gray."

"Yeah, I understand that we do not and never did, in fact, do what I do."

Gilmore choked on crab meat and paused to swill beer. "God, this is good. Want some?

"It's a mean underhanded world, Gray. You've done things an armored division couldn't have managed, but only four people know you exist, and only two of us know the details of what you've done."

Gilmore smiled when he added, "Believe it or not, I've had three immediate fully informed superiors over all those years, so that is the number of people plus me who know exactly what you've done." He laughed, "As a matter of interest, two of those men have died. Natural causes, incidentally.

"So, you won't be getting any more medals than I will, Gray. When your time comes your pension will about equal a Lieutenant Colonel's, but you will be just another clerk gofer put out to pasture."

Grayson said, "Well, I haven't thought closely about retiring, Brad, but if you are going, perhaps I had better pay some attention."

Gilmore nodded agreement, more vigorously than Grayson expected.

"Start thinking about it, Gray. You are too old now for what you do. You run on experience and skill, but would you climb the side of that Iranian building these days?"

"I've chosen your targets carefully for oh, maybe the last six years. You've gotten hard jobs that required smarts, not muscle."

"I've noticed."

"Anyway, I've a few things to say that I consider important, and they are the real reason for this meet.

"First is, I'm retiring to a town just below Annapolis, Maryland called Mayo. I'll be there if you need to see me. Not for business, of course. I'm out of that after this weekend."

"Mayo? I know the town, that's where . . ."

Gilmore completed the sentence. ". . . they make Gemini catamarans. Hell, Gray, you got me interested in multihulls when you first went to Saint Thomas for a year."

"I was down there for two years, Brad."

"I guess it was two years. You owe me for that vacation, Grayson. Anyway, I got interested, and I've owned a Gemini for a while now." Gilmore smiled almost hopefully. "Maybe we can sail together some after you are out. I'd like that."

"I would too, Brad. You've given me a life that I have enjoyed--as strange as that might sound in most quarters. I won't forget where you are."

Gilmore's head shook in wonder. "That you enjoyed! God, who else could ever feel that way about what you do?"

Gray's voice was normal. "It was you who said I had a talent, Brad."

"And I was unbelievably right. Yet, in all other things you are as regular as apple pie."

"I never married."

Gilmore looked a little guilty. "Gray, I never gave you a chance. I bought off three women that you seemed to be getting serious about."

"Three? Good, God, I suspected one, but there was no way to be sure."

"Once you are out, find a good woman, Gray, and marry her. Don't have kids. You are too old. Just marry and enjoy each other as your time runs down."

Gilmore's voice held deep sadness, so Gray ventured. "Did you lose your wife, Brad?"

His handler's sigh was heavy. "Passed on two years ago. Left a hell of a hole." He got himself together. "My daughter, Anne, over there, is married to a good man, and they have a son I'm fond of. They are my family these days, but it's not the way I had planned."

Gray could think of nothing to say. He waited for Gilmore.

"Now, I am to the main reason for this meet, Gray.

"My replacement is James Waverly. James N. Waverly, and he would be irritated if the middle initial was left out.

"Gray, the man isn't worth a piss hole in a snow bank."

Grayson was shocked by Gilmore's vehemence.

"Waverly's the new breed, Gray. He is Harvard MBA. He is a bottom liner, he is soft and self-serving. Waverly has friends, and he maneuvered his way. Watch him like a hawk, Gray. Do not depend on him. He is the kind who will waffle or leave you out to dry if the going gets really sticky.

"God, I bitched until I got tired of hearing myself, but all I got back was that the man deserved a chance. Jesus-damn, they make it sound like Little League Ball where everybody gets to play.

"I had no luck, Gray. I'm out, and Waverly is in. All I can do is warn you to keep your head up, and stick to our deal that you get to refuse assignments for your own reasons. Waverly has been told, and so has my boss. I've got to add that neither approves." Gilmore fixed Grayson with his eyes.

"You think about getting out, Gray. Your pension won't get any bigger by hanging on.

"There is a new crowd in there now, it's a new generation of business minded paper pushers. They don't go to the shooting range, Gray, they attend the ballet. They won't understand your work, and they will despise you for doing it.

"Keep all this in mind, Gray." Gilmore began cleaning his hands with Handy-wipes, and Gray guessed they were about done.

Gilmore rose and stuck out his hand. "Gray, I've respected you since I saw you marching into our first meeting. I've liked you almost as long. You are the kind of man I would have liked to have had for a son."

Gilmore grinned and slapped Gray's shoulder. "Although, I've got to admit that I would have moved a son into a different line of work."

+++

Waverly was a royal pain in the butt, but Gray worked around him. He made sure that Waverly was not part of any of his actions. That was usually easy because the new handler made as little contact as possible. Waverly clearly feared some horrible catastrophe into which he would be dragged and ultimately ruined. Waverly was the wrong man for the job.

The worst was Waverly's muttering about how the agency had to rid itself of all this clandestine murdering. Gray accepted that James N. Waverly reflected the agency's intended directions. It was then that Robert Grayson decided to make certain that no one within would dare make him a target for someone from without.

It was all relatively easy. Gray collected his weaponry and recorded his tapes. His money came in two batches.

The assignment was a drug lord, who of course never touched narcotics. The lord directed operations and deposited money.

Robert Grayson occupied a seat in first class. He and his target sat a row apart, otherwise the compartment was empty.

The drug lord was an American, a younger man who carried aboard a suitcase and an immense attaché case. He was booked on an overnight return from Nassau. Gray doubted the baggage contained clothing.

Well into the flight the man dozed. With the attendant absent, Gray rose, crossed the aisle, and stabbed a short, palm-held needle into the man's right carotid artery. The drug took barely seconds. The victim was still attempting to locate the sting of the needle when he began to sag. In an instant he was unconscious and appeared to sleep.

Now Gray had time. He withdrew a hypo of amyl nitrate from his own brief case. Properly injected, the massive dose would cause immediate death. Unless an autopsy was quickly performed, the drug would evaporate and diagnosis would be a terminal heart attack.

Gray injected the drug, not forcing the fluid. Summer was best because everyone wore short sleeve shirts, and veins were handy.

Gray removed the attaché case from beneath his victim's legs. He felt the body convulse, but the man was already unconscious and would feel nothing.

At his own seat, Gray opened his courier's large diplomatic pouch. He removed two sealing straps with single use seals. Each was clearly marked diplomatic material, property of the United States of America.

He placed the victim's attaché case and his needle kit inside the diplomatic pouch and snapped the lock. He swung the dead man's suitcase into the aisle. The diplomatic straps sealed the baggage and made it property of the United States. No customs, police, or foreign power ever challenged the sanctity of a diplomatic seal. Whatever the suitcase and the attaché case contained now belonged to Robert Grayson.

The passenger was found to be dead before arrival in the Bahamas. His nose wrinkling in distaste, the other first class passenger was quickly and apologetically removed to coach accommodations.

Robert Grayson, the courier, recognized and hailed familiarly by Nassau customs, was in his taxi en route to his hotel before the drug lord's body was removed from the aircraft.

As he supposed, the corpse's hand baggage was packed with money, all one hundred dollar bills, and all tightly compressed.

Grayson called a major Nassau bank and requested that a bank officer and an accountant visit his room.

Accustomed to such transactions, deposits requiring circumspection, the bankers counted Grayson's money with swift efficiency, created an account in his name and departed.

Gray sat on his balcony allowing the cooling breeze of a dying norther to soothe his slightly stunned senses. He, Robert Grayson, was now a millionaire. The dead man's colleagues might search for missing money, but they would find no trail, and if they somehow came against him, Gray intended to plant them beside their leader.

Gray's second and final financial acquisition was in diamonds. They were an unexpected trove worn in a criminal's money belt.

The hit had been deliberately Mafia type, two small caliber pistol bullets in the head. The victim was himself a successful killer skilled enough to be believed untouchable.

Laying over in his native Italy, en route to his home in Salerno, he had felt reasonably secure. He was not. Wearing a beard and sun glasses, Gray came in behind room service. He shot the hitman who was lounging at his television. He locked the nearly hysterical waiter in a closet by wedging a chair under the door latch.

On his way out, Gray noticed the money belt exposed by the body's fallen open robe. Why not? It would be in keeping with a mob hit. Gray used the fallen man's dining knife to cut the belt free.

There was money, a respectable amount. A successful killer returning home might possess such cash, but the diamonds--Gray had never examined jewels in quantity.

There were no small diamonds. These were multi-carroted gems. Each lay in its own soft leather pocket among folded over wallet-sized packets. He had dozens of the stones. Someone was going to be distressed.

Gray buried the jewels in a spot only he knew on his Pennsylvania farm. A dozen or so years down the road, he could begin discretely selling them. One in this country, another far away. Gray realized that if the gems were carefully marketed, he had once again become a millionaire.

12

Colonel Bolinger was in one of his moods, and Major Charlie Fobbs kept his head down and his eyes on his work.

Bolinger sat at his field desk tapping his fingers and studying his knuckles. Occasionally, he said, "Damn it to hell!"

Buff did not like having his nose shoved in it. He was a colonel, for God's sake. He commanded the cadre of a full regiment of patriotic fighting men. He sent two of his sergeants out to rough up one old guy and everything went to hell. Master Sergeant Dillard lay in a hospital fighting to stay alive, and Master Sergeant Crowder had a wrist sliced so badly that it could never be right.

The colonel saw only two bright spots in the situation. First, both men stuck to their claim that Grayson had assaulted them and did not implicate him or the militia. Second, no one but

Major Fobbs knew about the orders given Dillard and Crowder.

Still, there were damned few secrets in the county that did not eventually leak out. The Sergeants would explain to their wives, or drinking pals, and the word would spread that the militia had blown a simple operation. The colonel almost enjoyed his rage because he could feel his determination to square up growing and firming.

Bolinger had to let it grow because going the next step was BIG TIME. A commander could sit around and order a few ass kickings without much worry. Hell, that was normal living for some of his men. They argued, drank, fought, and shook hands. Killing a man was different. If caught . . . well, it could be bad news.

Bolinger needed just the right man, and he believed he had him, a soldier who could pick his moment and take out this Grayson without leaving a trace.

One shot would be the ticket. Without other evidence, police would conclude that a stray hunting bullet had dropped Grayson. It could happen, buildings and sometimes people were surprisingly often hit by stray bullets. People were always out in the woods with rifles. God, he made his living selling to hunters and gun nuts. If Grayson went down, everyone would shake their heads in irritation that someone had been careless and turn the page.

Colonel Bolinger had a trained man. When he had appointed Charlie Wolfe Regimental Sniper he had recognized the man's hunger to be a genuine war time, silent, long range executioner. Wolfe had attended two civilian

Sniper courses and had diplomas to prove his abilities.

Wolfe owned a special rifle and it was a good one. Buff Bolinger had himself glass bedded the McMillan stocked M700 Remington and installed the Leupold 3 1/2 x 10X scope. The Sniper regularly brought in targets showing three shot groups measuring under one inch at one hundred yards.

Wolfe and his chosen observer also had ghillie suits. Perfected by Scottish poachers, the ghillies provided exceptional personal concealment. The Snipers' ghillies were based on United States Marine Corps models and had camouflaged burlap strips tied into fish netting fastened onto their fatigue uniforms. The ghillies were absolute marvels, and properly employed, the militia Sniping team was virtually invisible.

Bolinger knew that Master Sergeant Charlie Wolfe was one of his good ones, and when they expanded into full strength, Wolfe would probably be a Company Commander. Bolinger had no doubt that Wolfe would do a perfect job.

If he had harbored doubts, Buff would have backed away. An important line was crossed when you ordered an execution, but it had to be done. When word began to spread that the militia might have instigated the Dillard and Crowder fiasco, people would then add, "And look what happened to that Grayson. He got shot, that's what." There would be no evidence, but people would wonder if there might be a connection, and the thought would generate that the County Militia might rate a lot more respect.

Bolinger decided to play this one very close to his vest. He said, "Major Fobbs, have Master

Sergeant Wolfe report to me at the knife fighting range."

Fobbs was gone with appropriate hustle, and Bolinger made his way to a heavily planked backstop where he sometimes scheduled knife throwing instruction. What a joke. Only in movies did hurled knives instantly bring down enemies, but his troops enjoyed the activity, and at short distances some were remarkably accurate.

The troops, about twenty this day, were listening to a talk across the compound, and Bolinger saw Wolfe rise and bring his rifle with him. The camouflaged piece was wolfe's trademark, and a good one for this operation.

They sat together on a plank bench, and Bolinger stressed the necessity for complete secrecy. He watched Wolfe's eyes glisten as the order developed. Good, the man wanted the job.

Wolfe made no argument. "I'll develop a plan, Colonel, and report directly to you within two days. I will speak to no one."

Because of the magnitude of the step, Bolinger needed reassurance. "You see the importance of this don't you, Sergeant Wolfe?"

The Sniper was clear and cold. "The regiment has been insulted, Colonel. I will make it even."

Bolinger watched him go. God, if he could have a full regiment of such men, even a company of such real soldiers, he could have an outfit that would make the Pittsburgh militia look like rabble.

Wolfe did not appear at the gun shop until the third day. He saw Bolinger alone, waiting patiently until no customers loitered.

"I have it worked out, Colonel. The shot will be routine, but I need my observer because we could have a long wait, and steady observing is damned tiring and could affect shooting."

Bolinger knew that. He would have preferred using only the Sniper, but Wolfe considered himself an expert and a professional. He would not appreciate being told how to do his job.

"You will want Master Sergeant Tilson, I assume."

Of course he would, Wolfe and Ray Tilson trained as a team. Bolinger found his palms dampen and there was strange buzzing in his ears. My God, they were going to do it. He heard himself say, "All right, but Wolfe--you orient your Assistant Team Leader." Wolfe liked that kind of military jargon.

Bolinger added, "When you have completed your mission bring your rifle here, and I will turn on a new Hart best quality barrel and rebed the action." That sat well with the Sniper. A new barrel eliminated any possibility of matching the killing bullet to the gun, and that pleased them both.

Sniper Wolfe explained the mission to Ray Tilson who repeatedly licked his lips and exclaimed, "Hot Damn!"

Wolfe said, "I scouted around Grayson's place. He lives way out on a farm. I thought his road might be the best ambush, but when I got to looking, I found he has a lot of windows, and one side of his house looks out across a field.

"Man, Ray, I found a hide that won't quit. We have a covered approach, and we'll be shooting from only a little over a hundred yards.

"The best way will be to have a look in daylight. We can wait in town until we see him drive out. Then we 'll whip up a dirt road I found and go in quiet to look the place over. We'll wear our ghillies but leave the rifle. There probably won't be anyone around, but it will be good training.

"We'll come back in the evening and probably make the actual shot after he puts his lights on. Of course, he might come out and sit on his porch, and that would be perfect. If he doesn't show himself for an outside shot, I'll get him through his window."

Wolfe was pleased with the idea. "We've practiced a lot of glass shooting, and at one hundred yards a match bullet will go through window glass like it wasn't there."

Master Sergeant Tilson said, "Hot damn!"

Gray studied his field alarm thoughtfully. It had been tripped for the third time in two days. He had been absent the first two alertings. When he had checked from the attic, there had been nothing.

He again reset the motion detector and waited. For a long moment the green all clear held, then the red warning again lit up. Gray headed for his stairs.

The sensor was old, perhaps it had begun to fail. Once a deer had chosen to bed on the grassy hide. The doe had a fawn which had slept nearby. Gray had enjoyed watching them . He could have another deer, but . . .

Gray raised the roof tile with his usual care, no swift movement was the ticket. He chose binoculars for a first look. Gray focused in on the hide.

Well what do you know! Gray's pulse did not jump. It never did. Somehow his brain assimilated the surprise and did not shoot adrenaline through his system. Emotionally he was affected about like noticing an interesting window display.

There they were, two ghillie clad figures right out of the manual, spotting scope and bipodded rifle, both camouflaged, of course, damned near motionless and looking hard. They had seen him come in, and they would be trying to see into the house. No chance, the interior was darker than the out-of-doors.

The question was, were they militiamen or boy scouts, or some kind of wannabes out practicing skills, or were they kill minded Snipers? Someone from his past? Possible, but he had tried to close those possibilities. The locals? More likely, but what was their intent?

Gray wondered if the right response would be to slide out a tunnel and jump them from behind. His binoculars searched the woods, but the two were alone.

That would not be the way. Questioned, they would swear they intended no harm and were just practicing. There was a better way.

Using pillows and rolled blankets, Gray made up a dummy in a white shirt and propped it in a chair near a window. He arranged the lamp so that his crude head and face would be in shadow.

Wearing a Harley-Davidson bandanna over his hair, Gray walked the length of his porch. He kept moving, but jerkily, making it impractical for a Sniper to shoot. He still felt better back inside.

The shooter had seen the bandanna. Gray placed it on his dummy. The Snipers could not ignore the connection. Gray judged the lateness of the day and put his lamp on a timer. The lamp would go on after full dark. Until then, the Robert Grayson dummy would wait in darkness.

Gray turned on other lights. They would keep the Snipers working. He went upstairs to prepare his welcome.

Gray removed the caps from his explosive wires, positioned the generator/detonator, and clipped on the bared wire ends.

He sat in the dark of the attic, watching the daylight fade, waiting to discover if the Snipers played at dangerous games or were killers for real.

His experience declared them the real thing. Guys practicing and playing at soldier would probably have packed it in and been home for supper. These guys hung on. They wanted a living target.

It became dark. From his post, Gray could not see his chair light go on. His signal would be the flash and bang of the Sniper's rifle. In the darkness, Gray's hands touched the generator making sure of its readiness.

They had been waiting a long time, but except for one trip across his porch, Grayson had not been seen. Wolfe had tried tracking the moving figure, but the shot would have been impractical. Wolfe's scope was on ten power, and at that high magnification the target jumped around something awful. For a moving target at one hundred yards he should have been on four power. Too much magnification could hurt.

Wolfe had not intended to shoot, anyway. He believed in one shot, one kill. He would wait until Grayson was still. If it did not happen tonight, it would happen tomorrow. He had time.

When the natuaral light dimmed, Wolfe lowered his magnification to four power. His scope would transfer more light at the lower magnification, and four power was plenty. Through the sight, Grayson would look as if he were twenty-five or thirty yards away. Wolfe could put the 168 grain Sierra match grade boattail in his target's ear, if he chose.

A light came on, and Tilson said, "Hot damn, there he is."

Wolfe heard his observer's excitement, but he kept himself calm. Grayson was partly shadowed by the lamp's shade, but the head shot would be easy. From their position, the glass window was straight on, and Grayson looked to be about two or three feet beyond the glass. A piece of cake, a dream shot, Wolfe stilled his breathing, concentrated on his cross hairs and began his squeeze. Only a corner of his mind heard Tilson's "Hot Damn."

The jet of the muzzle flash accompanied the blast of the shot. Gray thought he heard the bullet strike the glass. Any doubt he had was gone with the shot. Gray gripped his generator and twisted the handle, and for certainty, he gave it a second vigorous twist.

Across the field, a gout of fire erupted, as if a fire cracker in a pan had tried to lift a heavy lid. The thump of the explosion was no more noticeable than had been the single shot. Gray closed his tile and went below.

Wearing black with his face sooted, Grayson slid from a tunnel and made a circuitous route around the field and through the woods. The moon began its rise, and he could move decently.

Behind him the house was buttoned with the steel shutters drawn. Ahead, Gray expected, lay two dead men.

Gray found their transportation, an old truck. In the bed a sturdy rifle box was covered by an old tarp.

Protected by their ghillies, the Sniping team was hard to see. Gray dragged them deeper into the woods and used his red lensed flashlight for examination.

Death had been by concussion, and it had been instantaneous. They had not bled significantly, but their bodies were loose and bent readily in wrong directions.

Gray emptied their pockets and found a man's usual stuff. Professionals would probably have carried nothing. Gray hoisted them one at a time and carried them to their truck. He unceremoniously dumped each into the truck bed and covered them with the tarp. The rifle box was top quality, and he took it back to the Sniping hide to lay with the rifle, spotting scope, and ID.

The truck started, and Gray drove down the dirt road, onto the local paved road, and on to the highway. He went north for two miles, then off and into the hills. He parked the truck well within an abandoned quarry. Just for the hell of it, he punctured the tires, raised the hood, and ripped out wires. Unless someone looked closely, the truck would resemble an abandoned wreck.

Gray washed away his makeup in a rivulet and began his return along mountain trails. It had taken less than a half hour to dispose of the truck. To get home unseen using only moonlight required more than two hours.

The rifle was nice, but the finish had taken some beating in the explosion. Gray noted that the serial number had already been ground away. He locked the Sniper's armament in his secret room.

The ID was interesting. He had names and addresses. He also had County Militia identification. Both were Master Sergeants, and one had penned in before laminating: *Regimental Scout/Sniper.*

Gray memorized then burned all of the papers. He stirred the ashes in a can before dumping them into the river.

He was returning from the river when the rain began. From within the buttoned up house he would not have heard it. Gray laughed aloud. Perfect, no tracks and no scent trail, not that he expected any remarkable detecting by law enforcement. If the bodies were more than a few hours old when discovered, investigations would be mostly forensic and questioning.

Someone else surely knew the Snipers' activities, but Gray doubted that the police would be told anything. How could the militia announce that their ghillie clad Snipers were trying to shoot Robert Grayson dead. Not likely.

So, he had ignited the wrath of the County Militia. Gray almost grinned. Whoever had laid on this night's activity would be greatly sobered. Their fun was sure as hell over.

Had Grayson killed their men? Blown them up? Weird! How could he have done that? And, what to do about it? Try again? Very dangerous, and who would volunteer?

Gray went to bed at two in the morning with a reminder to himself to find out more about the local soldier boys.

If there were more out there with the Sniper team's skills, Gray had problems. His trap had worked, but determined killers could be very difficult. Grayson had firsthand knowledge.

That would be for later, Gray slept with his usual sense of comfortable satisfaction.

13

Bucky Larson called to ask about bringing his mother over. Gray said, "Sure." What else could he tell the determined young devil.

God, he made one friend and the miserable pup was a matchmaker with an eligible mother. Not much chance, Gray expected.

Gray had been up early. He had examined his window, and as soon as the business opened he ordered a new glass panel. The match grade .308 bullet had penetrated the outer glass, and seriously frosted and cracked the inner pane.

The deformed bullet was wedged between the panes. The jacket was flattened and curled back on itself. The bullet's boattail had crushed forward, and Gray doubted anyone could make an accurate comparison test, even if they knew they had the right gun.

He had dumped a shovel, a rake, wire cutters, and a claw hammer into his wheel barrow and skirted his field to the Sniper's hide.

The explosion had ripped the hell out of things. There were three distinct cavities, and flying ground had torn away leaves and broken branches.

Gray found his explosive's wire ends and clipped them as short as possible. He raked a large area and shoveled the loose material into his barrow.

He removed what was left of his sensor and used the claw hammer to pull the staples holding the wire to the tree. That wire he also clipped short.

He leveled the ground, ending up with a slight declivity where he had before had a raised platform.

Gray carried forest flooring from within the woods and soon had the explosion reasonably disguised. The area still looked too barren so he dug out pine and gum saplings and replanted them on the former hide. If anyone noticed his work, his explanation for all the digging and spreading would be the tree planting. No one, he would say, wanted a hole in his otherwise even and regular woods line.

Not that anyone would notice or comment. His guess was that the Snipers had scouted the place, and only they knew their hide. The impending kill would not have been bandied about even within the militia, and that set Gray to wondering just who in that organization did know.

The Snipers could have been seeking their own revenge for Bags Dillard and Fudge Crowder, but men like Dillard and Crowder were not the kind to inspire much loyalty.

Gray's experience told him that orders had been issued--and probably hungrily accepted. There were wannabes all over the country these days. Gray shrugged; now there were two less.

Grayson disposed of the explosion leavings, hosed out the wheel barrow, and put the cleaned tools away. He showered and shaved with more than usual care. He did not want Bucky's mother to see him as a bummy old recluse who slobbed around on a motorcycle, a guy who failed to say much about his past and whose current fame was slicing and stabbing two local clods--just the kind of bozo a nice woman longed to meet.

Gray had the front door open, and Bucky pounded on the frame and hollered, "Hello?" Before he got there, Gray heard his enthusiastic describing of the fortress-like walls.

Gray's first impression was, now there is an attractive woman. There was a squareness about her features that he liked. Firm, decisive, he might have decided.

Her eyes focused on him, her mouth fell open, and she audibly gasped. Her recovery was quick, and covered by an attractive hiccuppy chuckle, but her eyes still held astonishment, and Gray doubted that she heard her son's introduction.

Gray heard the words. "Mom, this is my friend, Robert Grayson. Gray, this is my mother, Anne Larson."

Gray took the offered hand, knowing that his own expression disclosed nothing more than decent interest. Gray liked that.

He gave, or tried to give Anne Larson the expected house tour. It was difficult because her son kept butting in, showing her things and displaying great enthusiasm.

The bullet suspended within the glass panes was an instant hit. Officer Larson was immediately serious, backtracking with his eyes, lining up the bullet's probable flight.

He said, "The shot had to come in under the porch eve, Gray, so he couldn't have been too far. We could run a line from the hole in the porch screen to the window hole and know pretty well where he shot from."

Gray was not encouraging. "Yeah, we could, Bucky, but what for? Whoever did it is long gone. Chances are he never even wondered where his bullet ended up." Gray had been insistent to the manufacturer that the window replacement arrive today. As a very special customer, Gray expected a truck with the new window was already en route. Gray wanted the holed window to disappear.

Bucky said, "What are you going to do with the old window, Gray? It would make a great coffee table top with the hole pointing down and the bullet showing."

"I think the company wants the window to show around, Buck. They're giving me a discount for the trade-in."

Anne Larson's was an expressive face, and twice during the window talk Gray caught her studying him speculatively. Once she was even smiling in a secretive knowing fashion that put his guard up. He was not sure what was in Anne Larson's mind, but Gray doubted that his personal attractiveness initiated the unusual and obvious interest.

At the loading bench, Bucky's mother demonstrated familiarity. She understood hand loading, and Bucky said, "I told you my mom knew guns, Gray, and, boy, can she shoot."

Grayson said, "I can see that she knows, Buck."

"Mom taught me most of what I know. The academy just added on."

"Did your dad shoot much, Bucky?" Gray's tone was casual.

Anne Larson answered, "Paul did not like guns, but I started young loading and shooting with my dad." Again Gray caught both amusement and speculation. What the hell was he dealing with here, some kind of trained operative?

The thought was dumb. Anne Larson had about as much skill at deception as a girl scout. Bucky would have said something about that kind of background anyway.

They were on the front porch using the old rockers when Gray asked, "Bucky tells me you moved into the county after your husband died, Anne. How did that happen?" Gray liked the easy way they fell into the familiar names. Anne Larson was nice to be around.

Again Gray found her smile knowing. "My dad owned the house we are living in, Gray. He's had it as a rental for a long time. Dad arranged for my job with Doctor Fritz. I am sort of his "Man Friday." I tend to the business while he doctors. It's a good job for me, and the first work I have had since college.

"Bucky finished at the university and got on with the State Police. It works out well for both of us."

Gray turned to the young officer. "No gal on the string, Buck? I see some pretty ones around."

Larson grinned. "As a matter of fact there is. Name is Kelley Fry, and we are both serious. Kelley works in appointments at the sub station, so we are in the same business."

Anne waxed enthusiastic about Miss Kelley Fry. The talk was family and comforting to Gray who could scarcely remember his own family years.

Bucky said, "I've got to take Kelley to meet grandpa real soon. He would be hurt if was left out."

Anne smiled, "You should take Gray down with you, Bucky. I think he would really like your grandpa."

Gray wondered if he had just taken some kind of shot about being old or something. He was failing to catch Anne Larson's drift.

On their way out, Anne handed Gray her business card and said, "Call me, if you have time, Gray." Her eyes snapped interest and promise that Gray was pleased Bucky did not see.

He heard himself choking something about calling soon. What in hell was the matter with him? He automatically searched his feelings. Pulse normal, respiration unchanged, clarity of thought--wandering. Geez!

Gray glanced at Anne's card.

Anne G. Larson
Attorney
Address and two phone numbers,

one home, one office

Small town stuff. No city lawyer would let his personal phone number fall into strange hands.

So, Anne was a lawyer. He had imagined her college education to be a Shippensburg BA with the degree in Primary Education or something similar. Somewhere along he had discovered that Paul Larson had also been an attorney. So, Anne the lawyer had stayed home and raised the boy while dad had brought in the bread. That was the right way, but it was certainly rare these days. Now both adults in a family worked like beavers just trying to stay even with the economy. The old way was better.

Gray placed the card near his phone where it would be handy.

Gray was pumping iron late in the day when it all fell together. He said aloud, "Holy hell!" then, "I'll be damned!"

He sat on his pressing bench mopping sweat and filling in holes. After a while he grinned, first wryly, then hugely. He slapped the bench and said, "That old pirate."

Gray's voice betrayed both astonishment and amusement, and for Robert Grayson that was unusual.

+++

Buff Bolinger was not one of the first notified. A gun shop regular had the information.

"God, Buff, did you hear about old man Bradley finding Charlie Wolfe and Ray Tilson?"

Bolinger's guts turned hollow.

"No, what about them?"

"Both dead in the bed of Ray's pickup."

"Dead? Both of them?"

"Yep, blowed up is what I heard." The man reveled in having the story.

"Blown up? What do you mean, blown up?" Bolinger feared to hear, but when Wolfe had not checked in he had known that something had gone wrong.

The news carrier shrugged, a little disappointed that he lacked details.

"Man I talked to didn't know exactly, and they'd been dead for a couple of days. About all he knew was that old Bradley was out walkin' his land and saw a truck sittin' in his quarry. Had the hood up and the tires were flat. Figured it was stripped and abandoned, but when he got close his dog turned and ran off.

The storyteller actually licked his lips. "Then, Bradley got a whiff of 'em. Held his nose, my friend said, and went for a look.

"Until he saw human feet, Bradley figured it was a dead dog or maybe a deer under the tarp. He hauled out and called the cops."

More details came in during the day. Gun shop customers came in to talk as much as anything else.

One man said, "Dave Willowby, from up the river, you know him, Buff? Well, Dave was there when they moved the bodies. He said they both had on camouflaged suits like they'd been bow hunting or something."

A listener exclaimed, "There ain't nothin' in season, why'd they be bow huntin?"

"I'd agree with that. If they was deer poaching, they'd use a rifle. There weren't no bows or guns in the truck, so that's guessing, but they sure as hell didn't crawl under that tarp by themselves. Somebody caught 'em doing something is my guess."

Bolinger did a lot of listening. It had to be Grayson. But blown up? Bolinger could imagine a shooting, but . . .

In the evening a customer who had a friend near the coroner said, "They say those boys were just bags of jello. Everything inside was blown loose, but their skins weren't hardly torn. Nobody can figure out what was done to them.

"Now who around here would do such a thing? Hell, stealing you might shoot 'em for, I could see that, but dynamite 'em or something?"

Inevitably, someone asked, "Say, Buff, weren't those guys in your militia outfit?"

By the time he closed, others had asked, and a few militiamen had called in searching for information. Bolinger tired of explaining that Wolfe and Tilson had been militiamen, but that he had no idea what they had been doing.

Police investigators arrived the next day. They knew about ghillie suits and facial camouflage. Bolinger explained that the militia was not on maneuvers and had not been holding a meeting. He had no idea what the dead men had been doing, but it had nothing to do with the militia.

Even a reporter came. Bolinger finally asked if the newspaper had questioned the dead men's minister? The dead citizens, Bolinger claimed, probably spent more time in church than they did at once a month militia meetings.

Bolinger sweat more than a little. Somehow, Grayson had gotten the Sniping team. Had he forced information from them before killing?

If he had, Grayson was revealing nothing to the police. If Grayson knew anything, it would have to be that Colonel Bolinger had ordered the hit. Bolinger's sleep was restless.

14

Gray read Anne Larson's telephone number from her business card and punched it in. He got a secretary who buzzed Mrs. Larson.

Gray wasted no words. "I'm going down to Maryland tomorrow and coming back the next day. Do you want to ride along?"

There was no hesitation. "What time will you pick me up?"

"Seven thirty sound too early?"

"Nope, I'll be ready. We can eat breakfast on the way."

Gray drove his old car, and they spoke of unimportant things.

Anne asked, "Why do you drive this old auto? What year is it, anyway?"

"It's a 1983, and as near as I can judge, it does everything a new model can. Why? Because I own it, and it runs well. What else counts?"

"You've got me there. If I mention good looks, I am shallow." Her laughter was contagious.

Anne took over before the Baltimore beltway. "Let me try this old clunker."

Gray immediately became drowsy.

Anne said, "Go ahead and doze off. We aren't overheating." Her voice tickled him. "If you had a newer car your seat would go all the way back."

Gray demonstrated that his did, too.

Her, "Damn!" was soothing.

Anne drove into the yard overlooking the short tidal creek that took a single bend before opening into the Chesapeake. She touched the horn and got out.

The old man met them at the porch steps. His grin was mischievous and victorious.

"By God, you did discover each other."

Anne said, "You old conniver, " and went for a hug.

Gray waited for the handshake. The old man's eyes sparkled with the craftiness Gray remembered.

Bruce Gilmore said, "It's good to see you, Gray. Real good." He hung onto the handshake longer than usual.

"So, you've discovered my daughter." His voice was expectant, and he held Anne in the crook of an arm.

"It took just about as long as I figured it would. Do you like each other?"

"Dad," Anne's voice held amusement.

"Ah hah, I knew it." Gilmore was hugely pleased with himself.

Brad Gilmore had a friend. Patricia Green did not live in, but she was clearly familiar with everything. Her first words were, "So, you are Gray. Brad speaks of you."

They sat in the catamaran's huge cockpit, feeling the tiny bump of passing wavelets, and Gilmore was pleased to explain.

He was surprised that his grandson, Bucky, had been the catalyst that brought the two together. "I always knew that boy had good sense."

Gilmore had planned long. "You know, Gray, I've thought of how you and Anne would be a perfect match since she got old enough, but for reasons we both know, but that these ladies never will, that was not practical." He turned to Pat and Anne. "Secret professional stuff."

Anne rolled her eyes. "I've heard that all of my life. Huh, I know more than he thinks, just like I've heard about Gray for the last thirty years."

Gilmore was offended. "I never said a word."

"Yes, you did, Dad. You would mention your best man, once you said Gray liked to sail, and there were dozens of other things that I knew had to be the same man. You stopped mentioning him a few years ago, and I had sort of forgotten. I didn't put it together until I saw him in his hallway. He fit the picture you had unknowingly painted, and then I remembered him from the single time I saw him at a crabhouse.

"His house, his looks, and the nickname just jelled." She sighed maliciously. "But Gray did not know until yesterday."

Gray laughed, "No, I didn't, but why would I? I never got a clear look at your dad's daughter, she was married, and I heard her first name only once all those years ago."

Gilmore explained, his voice satisfied, "Well, Paul died, and I was real concerned about Anne's future. Of course, the urge was to take her and Bucky in here, but that isn't making a future. That's just comforting an old buzzard.

"Anyway, I got to thinking about you, Gray, and the fact that you were, ah . . ." Gilmore searched for neutral words. "well, leaving government service and retiring. I still thought you two should have a chance to know each other. The time was right, and I had an angle.

"Way back when you bought your land, I came up for a look, nice area, small town America, a place that would grow, but not too much. I bought the house Anne lives in for pocket change, rented it out, and let the investment grow."

Brad looked even more satisfied. "The rent always paid the taxes and mortgage. Now the house is paid off, it's worth something, and best of all, it fitted into my scheme.

"All I did was arrange for Doctor Fritz to give Anne a try. That worked out, of course. I booted out the tenants, had to pay them a bundle to go quietly," the old man grumbled, "Then I sat back to let things develop."

He grinned widely, enjoying his scheming. "It took so long I was about ready to jog one of you, but it was best to discover each other."

As if savoring a rich desert, Gilmore said, "Tell me how you found out who Anne was, Gray." He clearly hoped for intrigue.

Grayson went along with it. "It was obvious from the moment I saw her that something was amiss, but I couldn't work it out. I read her business card, but the middle initial "G" did not mean anything.

"Then, like it so often does, everything just fell into place, and I knew. Too many coincidences, I suppose. Bucky spoke about the Chesapeake, Anne kept looking funny at me. Hell, you never know what makes it all clear. Anne knew about handloading, and Bucky said she had learned pistol shooting from her grandfather. There were a million indicators, once I found the direction.

When I got the hunch, I contacted a friend who has all the driver's licenses in the United States on CD ROM. I got Anne's middle name. All vehicle registrations have full names. Gilmore. Then I was sure."

Brad sat back with a sigh. "It is remarkable what fate can do if you give it a little shove. Of course, I was keeping watch, just in case either of you got another interest. I'd probably have had to offer the "other" a super job in the Swahili desert."

The wind was right, so they sailed. With a full crew, they pushed from the dock and short tacked out the creek. They reached up and down the bay enjoying the catamaran's swift and flat ride. After an hour they dropped the main sail and downwinded into the creek under the jib alone. Brad brought the Gemini across the wind ,and Anne popped the jib sheet at the exact moment to let the boat ease alongside their dock. Gray cleatéd a spring line, and they were in.

Gilmore said, "Gray, that's dream sailing. Nothing like a good catamaran."

"I can't argue, Brad, but it's got to be a good one. Some of the roomarans they are popping out sail like log jams. Horrible."

Gilmore was insistent on tying his boat off his way. Gray stayed to watch and to have his friend alone for a few minutes.

Brad said, "You do like Anne, don't you, Gray?"

"For God's sake, Brad, this is the second time I've talked with her. Anyway, she may not like me."

Gilmore's chuckle was knowing. "Oh she likes you, Gray. Would she have come down here with you if she didn't?" He almost he-hawed. "Would you have invited her if you didn't?"

Gray muttered, "You're a real detective, Gilmore."

Gray said, "Sit down a minute, Brad. I've got something serious to talk over."

Gilmore recognized the tone and became professional.

Gray began with the motorcycle theft.

GIlmore said, "You just snapped his neck? Whew, you don't fool around, Gray." There was no condemnation. They had shared a hard vocation.

Gray nodded, unbothered, and went on. He told about the cafe attack and his work with the steak knife. Gilmore whistled, and his eyebrows rose.

Gilmore was increasingly surprised. "You blew them up? Why not question them and find out what in hell was going on?"

"Jesus, Brad, playing Indian in a dark woods is chancy stuff. It doesn't always go one way, you know. I had to let them take their shot to make sure they were serious bad guys. In the dark, I couldn't shoot, and they would as sure as hell have hauled out of there within the next few seconds."

"So, how do you figure it, Gray?" Gilmore did not wait for an answer. "It's almost got to be the militia outfit. It's too military and too terminal for just a couple of good old boys getting even for a cutting."

"That's the way I see it, but how in the devil can I find out who knows what, in an outfit like that?"

Gilmore thought about it before advising.

"There is no practical way to infiltrate. Hell, you are a stranger in town. Nobody will talk to you, but there is the old, rarely failing rule to consider. 'Cut off the head and the body dies.'

"Most rag-tag terrorist groups have one charismatic leader that holds them together. Hell, that's true of most athletic teams and social clubs. Your militia is probably the same. They will have one man who keeps them going. Remove him and the rest will melt away. It almost always works."

Gray knew the rule. "Sometimes, the key man stays in the background, pulling strings and letting someone else lead. That could be the case here."

Gilmore watched a crab working around a piling. "You will want to be sure, Gray. A string of hits when you are already involved would be poor planning. Get the right man the first time. That was always our rule."

Gray recognized the strangeness of their talk. They spoke of killing as they might discuss a menu. Neither gave an instant's consideration to normal judicial channels. Brad Gilmore had ordered killings for forty years. Gray could wonder how many. He had not been Gilmore's only agent. Probably his favorite, but not his only executioner.

Gray? Gray tried not to keep count, but there had been a couple most years, and he also had worked at his trade for decades.

Neither man flinched at ending a life, Gilmore through a career of such ordering, Grayson because he felt nothing. Gray ate, he slept, he . . . killed. Each act was natural, and the eating and sleeping affected him no less or no more than the killing. Unnatural? Obviously, but neither man thought about that anymore; they hadn't for thirty years.

Gray said, "Bucky knows something about the militia. He may have the whole picture. I'll check with him first."

Gilmore rose and patted his boat's wheel affectionately. "Keep me informed, Gray. I like to stay up on things."

He stepped onto the dock, and Gray asked, "Do you want to hose her off?"

"Hell, no. A boat is for using not working on."

Gray grinned, "That means you've got someone hired to do it for you. Probably some young sailor that you pay by letting him use the boat."

Gilmore looked surprised. "Of course, Gray. How else would you do it?"

Brad said, "Let's go up. We're having crabs."

Gray groaned, "Brad, a man could starve to death eating crabs. You work yourself weak peeling the damned things, and all you get are little dibs and dabs of eating. I'd rather get mine out of a can."

"Heresy! You are a barbarian, Grayson."

"Maybe, but I won't starve to death. God, eating dead crabs that have been eating other dead things on the bottom of a polluted bay. That's disgusting."

15

Colonel Bolinger needed to discuss his plan with someone. He needed support. Even the greatest leaders had their staffs and trustworthy aides. In this case, Adjutant Charlie Fobbs would have to do.

Who else was there, really? Besides himself, only Fobbs knew the militia's inner workings. Fobbs was not worth a hoot in hell as a leader, but he was loyal and did damned good staff work. The rest, all those captains and lieutenants? Bolinger was not about to include them in a plan as sensitive and as daring as this one. Nope, he and Charlie would have to do it alone, all the procuring, mixing, and driving.

Provided he decided to go ahead, of course, and Buff was not exactly sure, just yet. It was one thing ordering. Quite another doing a deed himself. This one would be a sweaty business with awesome repercussions.

Bolinger found himself vacillating, and that embarrassed him. He had always preached doing what had to done, no matter how difficult or dangerous--hard, all man, warrior talk. Talking, the colonel was realizing, was more comfortable than doing.

A communiqué arrived from the Northern Pittsburgh Area Militia. General Bill "Guts" Blakey wrote personally. In part he said, *"Somebody's kicking your asses all over Eastern Pennsylvania. Hell, we count four of your boys down plus one of ours. Seems to us out here that you all ought to take hold and inflict a few casualties on whatever crowd is working on you."*

Bolinger agreed, but he had already sent his best, hell, his only team. Finding a foolproof way to take out a man who was obviously an experienced fighter did not come easily. Grayson, and that was who Pittsburgh was talking about, had handled Bags and Fudge without getting out of his chair. The Sniping team was a mystery, but it had to be Grayson. No one else was involved.

Close questioning by police investigators gave Bolinger his plan. Their insistence on examining the militia's pathetic armory and ammo dump was both mortifying and revealing. Bolinger could have resisted and insisted on proper search warrants, but the regiment had nothing physical to hide. In fact, they had damned near nothing at all. Militiamen brought their own rifles and ammunition. That had been true since Lexington and Concord. Buff had reminded the cops of that historical detail.

The investigators were interested in explosives. The regiment had once owned three

World War Two hand grenades, but they had been expended in training. The militia's larder consisted of three bolt action, .22 caliber rifles and two bricks of ammunition used for basic marksmanship training, more than a little embarrassing, Colonel Bolinger felt. He hoped word of the regiment's poverty did not circulate.

Buff got Charlie Fobbs alone in the gun shop and explained how things were.

"Grayson blew up Wolfe and Tilson, Charlie."

"He did? How do you know that, Buff?"

"I know it because I sent our Sniper team to get Grayson."

"You did?" Fobbs gulped down sudden and gut wrenching anxiety, and his eyes rolled a little while his methodical mind digested the information.

"That's why they were wearing their ghillie suits, Charlie."

"But their rifle wasn't there, so maybe . . ."

"I called Wolfe's girl. The rifle is gone, and so is the spotting scope."

"She might tell the police that you asked, Buff." Fobbs' voice was panicky.

"No, Charlie," Bolinger was patient. "I own this gunshop, remember? I explained that I might want to buy Wolfe's guns. She read them off, and I will buy them, but the Sniping stuff is not there."

Bolinger gave his adjutant only a moment, then went on. "Grayson killed them, Major. Who else could it be? I don't know how he did it, but somehow he got close and blew them up. I've been trying to find out what kind of explosive was used and whether they were full of grenade fragments, but nobody will talk to me yet."

"This Grayson must be one bad dude, Buff. He cut Bags and Fudge without even raising up." Fobbs' shaky voice speculated. "Maybe he even cracked that Pittsburgh guy's neck and made it look like an accident."

Bolinger felt his neck hairs rise. He had not considered that possibility. If it were true, Grayson was super dangerous, and if he would kill a man for trying to steal a motorcycle, Grayson was not likely to be satisfied with just wiping out a pair of Snipers. Grayson would figure that the Sniping team had not gone out on their own. He would want whomever sent them. That made Buff's new plan all the more essential.

Bolinger worked his new thinking through Charlie Fobb's slow and jittery mind, then added, "We've got to get him before he gets us, Charlie."

Fobbs was horrified, still only half believing, and scared nearly witless. "You really think he'll guess who sent them, Buff?"

"He'll figure out that it was militia doing. He's probably thought that out already. That'll mean to him that the head militia officers are at fault. Yep, he'll come for us, you and me, Charlie, and he won't be slow starting. Like you said, this is a bad dude."

"What'll we do, Buff?" Charlie Fobbs was clearly a frightened man.

"What we'll do, Major, is we'll get him first, like I just finished saying. I've got a plan, but we'll have to do it, you and me. We can't let anybody else know, and it sure as hell has to be done exactly right."

Bolinger made his voice confident and certain sounding.

"First, we'll reconnoiter. You'll drive, 'cause that's what you'll be doing when it's for real. I'll do the rest."

They drove all the way to Grayson's house and turned back through the man's "U" shaped driveway. Bolinger kept saying, "Look close, Charlie, you'll be driving in the dark."

They took Buff's pickup for the materials gathering, and Major Fobbs became enthusiastic in a desperate and excitable way.

"Oh, I know about this, Buff. Man, we can make as much as we want. How big a bomb do you want, Buff?"

Bolinger explained. "We'll mix the fertilizer, fuel, and the other stuff tomorrow. In the afternoon we'll go and get Bags Dillard's big truck. It's parked beside his shack, and I know where the magnetic box with extra keys is stuck on. I'll bring it over here, and we'll load up. We'll keep the truck inside the barn until late night, then we'll go for it."

Bolinger bought a barrel of fertilizer and twenty miles later Fobbs bought one. They bought repeatedly until the truck was full. They were two counties away and in farming country where everybody bought barrels of fertilizer all the time

Mixing was pure hell. They poured, dumped, stirred, and shoveled. They had to rest a lot, but Bolinger kept reminding how they had to do it, and do it right now.

Charlie Fobbs was scared and kept raising irritating questions, but Bolinger was patient because answering helped make them both sure they were doing everything right.

"You sure we burned every receipt?"

"Every one, Charlie."

"Suppose he ain't home?"

"We'll wait until tomorrow."

"Will this stuff keep?"

"Yep, but if it takes too long, then we'll just burn this batch and make another."

"How'll we know if he's home?"

"Wolfe said he leaves his car outside."

"I heard his house is some kind of fort or something."

"That's why we're mixing so much."

"How much you figure we've got?"

"About all that pickup can hold."

"God, that's a three quarter ton truck. We can move a ton and a half on it."

"That's how I figure it. I'm planning on blowing his damned house into splinters and cement chunks, if that's what his fort is made of."

"I remember reading that one thousand pound bombs were called Blockbusters back in WW II because they would take out a whole city block."

"That's what I've heard."

"Good-by, Grayson."

They had mixed on an old concrete cow barn floor. Bolinger insisted that they scrub the place with detergent and get it clean. Then they redistributed manure and hay leavings until it looked natural.

Bolinger said, "Damn it, I still smell fuel oil."

"Hell, we're soaked with it, Buff."

"I guess that's it. We'll get out of these clothes and burn them. No traces, Charlie. If this gets tracked back to us, we'll never get out of the can."

Fobbs shuddered.

Loading Dillard's truck was a brute. When loaded, the rear springs were bottomed out and the hood stuck up in the air.

"Hell, Buff, you'll have to creep along. That load'll barely steer."

"Who cares, we're only going forty miles or so. I'll take it easy. You can go ahead and wait for me along Main Street, where you can see if Grayson passes, and where you won't be noticed.

They covered the barrels with an old tarp and were ready.

It had been a long, but pleasurable day. They had returned from the Chesapeake by noon.

Like some ancient Iroquois warrior, Gray had circled his property looking for enemy sign. Most of his work had been urban because the bad guys worth assigning him to were always rich, and few with wealth chose remote areas.

Most evil people lived in cities, Gray found, so he had made cities his business. If his county enemies were truly skilled, Gray might miss their sign, but few in the modern world were genuinely expert woodsmen.

It could be said that few violent criminals were ever expert at much of anything. The city street shooters were the most obvious examples. Organizations wishing to disarm ordinary citizens invariably claimed that the crook with a gun had a better chance of winning than an armed home owner. That the opposite was true, and that every study or survey ever conducted proved that armed citizens almost always took out their attackers swayed few.

The facts were that armed criminals do not practice. Criminals simply own a gun and a magazine of ammunition for it. Pistol, shotgun, and rifle wielding thugs are most dangerous to bystanders. Most criminals are unskilled and careless, but not all, so Gray remained cautious.

Later, Gray worked with his weights and hammered three rounds on the heavy bag. He showered and napped, then went to Anne Larson's for supper.

Anne did not put on a show. She set out both milk carton and plastic wrapped bread loaf, just as she would have for her own or Bucky's meal. The butter stick was partly used with knife marks at one end.

Gray appreciated the absence of pretense. He wished to discover the real Anne Larson, not a staged personality that would have to be worn away to find out who lived within.

Bucky came and departed before they settled in for conversation. Gray believed it essential to place certain cards in plain view. He had never done so before, but if, as Brad Gilmore obviously anticipated, there was to be serious exchange and closeness between he and Brad's daughter, a certain clarifying had to be accomplished.

When he could, Gray asked, "Do you know what your dad did in government service, Anne?"

"Not exactly. He never spoke directly about it, but over the years I put together a sort of picture of what he did."

"You know that he was my boss, of course."

"Oh yes."

"But you do not know what I did."

"I think I do, Gray. You were and still obviously are my father's favorite. I heard things, scraps of conversations, one side of a phone call, bits and pieces here and there. You were some sort of trouble shooter for Dad, weren't you?"

Gray was surprised by even that much.

"That would be a good description, and I should probably leave it there. My duties are classified by a need to know that is virtually impenetrable because every level of our government requires plausible deniability. What I did never happened, period."

She waited, and Grayson chose his words carefully, not wanting to horrify or frighten off, but sensing a need for honest field leveling.

"During the war, the Korean War in my case, I killed some of the enemy. It was noticed that I was able to be calm and deliberate where other soldiers became excited, panicked, or confused. You could say that I was fearless and that I could kill without even a hint of regret or hesitation. Those descriptions would be close enough. I had a special talent that our government needed, but could never openly admit.

"To make it short, I continued soldiering until I retired. Those who chose me were exactly right. I have never experienced hesitation or regret. I found my duties challenging and usually interesting."

Gray had not looked away. He closed by saying, "I thought you should have an idea of the kind of man your father is aiming at you."

Anne Larson's lips formed a straight line. "I thought it was something like that." Then she added, "But perhaps I do not really understand what you are saying, Gray. I may be reading something you do not mean."

Gray nodded, "Here is an exact example. This will be the only incident I ever expect to mention. I can tell you this because it was not exactly a part of my official duties.

"Many years ago in the heat of an explosive operation"--Gray was amused by his choice of words--"I smashed a woman unconscious with a hand grenade held in my fist. The blow could have killed her. She was a known enemy who was reaching for a pistol. She was also severely injured when a grenade I placed exploded near her.

"The woman was in enemy intelligence, and thereafter she devoted herself to uncovering the identity of the individual who had disfigured and almost killed her.

"When I retired, most of my protection ended. Not deliberate withdrawal you should understand, but names and details considered history could get passed about. The woman still hunted for me, and she offered attractive rewards for my identity.

"So," Gray paused, his mind visualizing, "I decided to get her before she could get me. I did exactly that and ended a threat that had gone beyond official and become fanatically personal.

The woman always served in the United States or England because she looked and sounded exactly right. Gray located her duty assignment before he retired. When he was ready, Gray flew to England via Madrid. He watched until the woman walked unescorted. Gray joined the pedestrian crush and worked close behind his quarry.

The dart was tiny, but its poison was deadly and quick acting. Death from the toxin was ugly with body convulsions that would keep volunteer helpers at bay until medical help was too late.

Gray blew the dart from a shortened plastic soda straw gripped within his fist. To watchers, a man coughed and politely covered his mouth. Gray aimed for the woman's spine between the shoulder blades where she could not reach the dart.

Gray coughed, and blew, and slowed his pace to examine a window display. After a moment, he resumed walking but in the direction from which he had come.

A half block behind him there was a disturbance with voices calling out. No one near Gray even noticed. He discarded the soda straw in a waste container.

There was also the Fat Man, who was not fat at all. He had a real name, but no one used it. He had been the woman's investigator and information source. Grayson hit the Fat Man only an hour after disposing of the man's sponsor.

He was called the Fat Man because of what he ate. Pork roasts, fat ham, and marbled beef were complemented by cheesy and mayonnaisey dishes. The Fat Man was actually thin and round shouldered with pointy features and a bald dome of a head. The man did have a protuberant paunch, and his trousers were cut to fit neatly, but had to be supported by suspenders.

The Fat Man was every man's information bank. For a fee he would search for anything. For a fee he would sell anything he discovered. For years he had requested information on the cat killer who had scaled the five story building and had all but killed the woman. To date, the search had fallen short, but . . . Gray could not be sure what the future held.

The Fat Man was tolerated because he could be useful to all sides. To kill the Fat Man was as serious a matter as destroying a universally owned library. The man seemed to know something about everything clandestine, and everything about many things.

Gray did not care. He had never used the Fat Man. Everyone might be better off if no one could use the man's gatherings, but Grayson did not care about that either. The Fat Man was in the way, a danger, so Gray would remove him.

Fat Man was a difficult target, but he ate daily at a side street restaurant. His vehicle appeared, stopped, and if the narrow sidewalk was clear, he exited and three-stepped into the protection of the establishment's entrance.

Gray's solution was a narrow alley walkway directly across from the restaurant. It ran for a block faced with almost windowless brick buildings, and was barely wide enough for a single cart's passage. Well along there was a bricked-up doorway where handcarts had once lined up for deliveries. The narrow loading dock remained. From that high and narrow perch, Gray could see over a motor car and into the restaurant's entrance. When the Fat Man scurried, Gray would have a clear shot at the man's back.

Grayson's weapon was an AR-7 with a full automatic conversion and a ten round magazine. A silencer lengthened a shortened barrel, and the wire stock folded away. A four power telescopic sight provided sighting. Gray carried the rifle in a violin case, a common enough sight in England, if unimaginably gangster-like in the United States.

The Fat Man was reasonably prompt, and Gray had not been waiting long. A black taxi pulled up, and Gray braced his rifle against the bricks. The Fat Man appeared. Gray's crosshairs settled on the back of his head where his coat collar ended. The man took his first step as Gray's trigger finger squeezed. He started his second as all ten standard velocity .22 caliber Long Rifle bullets arrived. The strike seemed almost simultaneous. At fifty yards, Gray's group was tight.

Gray dropped from his perch and left the untraceable rifle. In moments he was away.

The killing in a city street horrified Londoners, but those who knew the Fat Man wondered that it had not happened years before. Word spread that the IRA had executed an enemy.

To Anne Larson, Gray did not mention the Fat Man, and he did not describe the death of the woman. Those stories and details would remain forever his. Even Brad Gilmore rarely learned details. Except for Gray's wall climbing, Gilmore had never seen his man work. Robert Grayson did not have an "M" or a Moneypenny. Gray was never debriefed. That complete secrecy, both he and Gilmore agreed, was why Gray had lived to retire.

Anne Larson had grown up with intrigue and secretive service. Gray's revelation did not shock her. She nodded, accepting all of the unspoken assignments that must have made up a career. She said only, "That is about what I expected." And their evening continued.

Gray turned in late. With the Snipers in mind, he closed all of his steel shutters before turning on lights. He showered and welcomed his bed.

His last acts turned out his bed lamp and opened wide his bedroom shutter and window. Gray slept on the second floor where it would be difficult to reach him.

16

Colonel Buff Bolinger's hands shook on the overburdened truck's steering wheel. Bolinger was scared. He was scared into an almost dumb sluggishness that left him barely functioning.

A portion of his mind, somehow divorced from his terror, chastised his cringing soul. What in hell kind of a soldier was he, anyway? He had a rock solid plan without any kind of danger. If he got to Grayson's and everything was not right, he would simply "U" turn out of there. He was still scared. Gut sick, damp-palmed, leg-shaky scared.

He turned onto Grayson's half mile long dirt road almost surprised that his muscles responded. He had to leave his headlights burning, but he prayed that Charlie Fobbs had remembered to douse his going in. The plan was for Fobbs to go first. If anything at all was wrong, he would get his headlights back on, and Bolinger would see them. The road remained dark, so all must be well.

Bolinger slowed to a jogger's speed to be sure nothing went wrong on this last stretch. The white stucco of the house loomed suddenly, and there was Fobbs in his car just where he was supposed to be. Bolinger experienced a sudden surge of confidence. By God, they were going to do it. His right hand checked for the Bic lighter on the passenger's seat, and it was there. More confidence flowed. He sensed Fobbs pulling in behind him, but his mind was measuring distances.

The plan was as basic as it could be made. Industrial fertilizer was used as bomb material by miners all over the nation. Properly mixed and with a powerful detonating agent, anyone could manufacture the hugest bomb he could transport.

Bolinger's detonator was the simplest. He had stuffed a stick of dynamite inside a ten pound can of 3FG black powder. A blasting cap was pressed into the dynamite, and he had a one and a quarter minute fuse. Once lit, the detonating charge would be impossible to stop as it was buried within almost two tons of fertilizer explosive.

He was there. His truck grill faced the end of Grayson's long front porch. Heart thudding, hands again shaking, Bolinger fumbled the Bic lighter into fire. He held the flame to the end of his fuse and saw it catch and spark as burning began.

Instantly, he fed power to all four pickup wheels. The heavy truck rumbled ahead and with only a slight bump was onto Grayson's ground level porch. The light bar on the cab was crushed and ripped away by the porch overhang, but the truck plowed ahead.

Nearly halfway along the porch, Bolinger's driver's side mirror bent backwards from striking a bay window. Buff killed his engine, ripped loose the key, and crawled for the passenger's door.

In an almost panic he could not locate the inside door handle, but the window was open. He got the outside handle, pressed the button, and was free.

He had lost count of the seconds, but believed he was moving swiftly. Headlight reflection allowed swift movement back along the porch, and he saw Fobbs waiting in his blacked out car.

Bolinger again had trouble with Fobb's door but cursed himself inside. When the door opened the interior light went on. Fobbs said, "Oh God, I forgot."

Bolinger said, "Drive, damn it, drive."

Fobbs burned rubber, flipping on his headlights, and they tore away on Grayson's half mile long dirt road.

They reached the end and turned down hill. Safe! They were now just innocent passersby. Suddenly worried, Buff asked, "How long has it been?"

Fobbs' mouth opened to reply that he had no idea when the sky turned to daylight.

Bolinger said, "Yes!" Then the shock wave of the blast rocked them. Bolinger was awed by the power. Over a half mile away and the whole vehicle shuddered. Fobbs stayed on the road, remembering to slow to a normal speed. A few blocks later, he turned and pulled off into a cindered patch where railroad cars had once parked.

The sky glow had faded and no firelight reflected from the cloud cover.

Fobbs said, "Maybe it ain't going to burn, Colonel."

Bolinger's laugh was short and harsh. "It might be hard to find anything to burn. That damned house must be spread over three counties."

"Maybe they won't be able to find his body." Fobbs sounded hopeful.

Bolinger did not share his hope. "They'd better. If they don't, we'll always wonder if we got him, or if he's out there waiting for just the right moment to get us."

Firemen sped by, their personal red emergency lights showing. Fire trucks began their runs, and Bolinger and Fobbs could follow their routes by the lights and sirens.

Bolinger snickered. "And there's excitement in the old town tonight."

Fobbs started his engine, but Bolinger reached across and shut it down. "We'll wait, Charlie. A lot of people will be out looking around, and we don't need anybody remembering that we were in the neighborhood the night Grayson got blown up."

"The police will come to see us, won't they Buff?"

"They'll come to see me, Charlie. They'll tie this blast to two of our men getting blown up, and they'll ask questions. Hell, I don't know Grayson, never had contact with him, didn't know where he lived until he blew up.

"Figure it out, Pittsburgh is sore at us because their man got killed, so they blow up two of our people. Then they get Grayson, who, like you suggested, might have killed their liaison man. Us? Hell, Charlie, we're innocent bystanders."

Both men bellowed tension relieving guffaws.

Headlight glare and the roar of a vehicle engine woke Gray. A crash of breaking metal and glass forced momentary disorientation because the crush of sound was coming from a strange direction. The engine stopped, a vehicle door slammed, and Gray got himself together.

His second floor bedroom window opened onto the back gable end of the house, but Gray judged someone had driven right onto his porch. Gray's actions became instinctive. He snatched his .380 from the bedside stand and keeping low, he crawled to his window. He cranked closed the bulletproof glassed window and began rolling closed the steel shutter.

The house was otherwise buttoned up, and the car was probably some damned drunk wondering why window service was not open, but . . . Gray had almost closed the shutter when the blast came.

The explosion was beyond hearing. A monumental shock wave erupted and as always tore along paths of least resistance. The truck's body metal disintegrated. Melted and distorted, the frame sunk into the cement porch floor as if the concrete was merely mud. Engine, transmission, and wheels either blew into fragments or departed en masse for distant places.

The porch roof with its tons of tile only minutely contained the blast before being pulverized into unidentifiable particles. The surrounding heavy foliage from porch level up disappeared, and across the river windows were blown in and roof shingles tore free.

Grayson's wooden house with its monstrous bulletproof windows was instantly gone. The giant pressure wave struck the three foot thick reinforced concrete walls with nuclear force. The outer wall crushed into the inner, but held together. The massive steel window shutters directly against the explosion bowed inward until their tops freed allowing the already declining blast to roar through the house at velocities exceeding the worst of hurricanes.

Most of the explosive power was tipped upward by the resistance of the pillbox walls. The night lit up beyond the ability to watch, and the blast wave rolled across the flat box-like bunker roof to remove, destroy, and redistribute the field side of Grayson's home.

Although his window opened perpendicular to the explosion, the blast curled around the structure and a furnace heat burst through Grayson's partly open shutter. Only Gray's arm was above the window ledge, and it was protected by the steel shutter, which was struck at an angle and did not buckle.

Grayson knew nothing of any of this. He may have seen the light for an instant before the shock wave slammed then lifted his body. The impact of compressed air rushing through his window colliding with the greater force raging through the building from the failed downstairs shutters blew his bedroom door apart, hurled everything movable, and ripped sheathing from the walls.

Grayson was vaguely aware of a blindness and incredible buffeting, but his battered senses could not identify. His body slammed into something, and he began to lose consciousness.

The blast moved on creating an enormous vacuum. Air was sucked away, and in his semi-conscious condition, Grayson's lungs struggled to breath. Awareness fell before the air returned in a wall of powered pressure almost as damaging as the initial explosion. Again loose objects flew. Grayson's inert body was rolled, but he could not feel it.

Quiet returned. A few things fell, but the darkness was again undisturbed. Small fires had caught and burned fitfully outside the tortured structure. Within, the silence lay heavy.

The volunteer firemen were fast. Highly motivated, they did not spare themselves. Despite the lateness of the night, the first truck left its station only five minutes after the explosion. Within fifteen minutes everything rolling was en route and surrounding departments were alerted and standing by.

The first fire fighters attacked the small blazes and searched for victims. Until they located the blown-in shutters there seemed no way to enter the bunker. There was no appreciable smoke, no gas, and no fire inside, but the explosion had destroyed the shelter's interior.

They found Grayson on his bedroom floor just beginning to move. He gripped a semi-automatic pistol and the fire fighters backed away and called from beyond the room until he slid the gun across the floor and out of reach.

The paramedics of the volunteer ambulance company were second on the scene. Robert Grayson was their interest, and they tried to help. It was astounding to everyone, including Grayson, that not much help was needed.

Gray was still wondering where he was when firemen's lights began probing his upstairs. He was not surprised to find his pistol still gripped in his hand, and he had difficulty loosening his fingers to remove the pistol's threat so his rescuers dared enter.

He could not at first accept the destruction of his home. That his mighty bunker had been fractured testified to the giant size of the bomb that had been used against it.

The memory of headlights, an engine, and a vehicle door slamming came back. Someone had come for him big time.

Gray had not built to protect against such an awesome weapon. He sniffed and suspected he could detect the alfalfa scent of the explosive used.

He had survived, and survival in the most powerful way proved the success of his design, but he saw his steel and concrete bent, some melted, and more blasted into rubble. This was Beirut or some third world country's way of vengeance.

His rescuers wanted him out of the building, but Grayson held them at bay until they found his bedroom flashlight, his pants, and a pair of shoes. Gray pulled the trousers on over his pajama bottoms, cinched his belt, and despite protestations jammed the .380 into its holster in the small of his back. His shoes allowed movement on the debris littered floor, and he shook out a blanket to drape around his body. Gray was not fooled by his lack of visible wounds. He had been blasted, hurled, and nearly asphyxiated. He could be hurt internally or shock could set in. He wove a dizzy way to his staircase, stumbled down its shattered remains, and made his way outside.

The night was lit by dozens of car headlights all pointing at the explosion site and by the fire department's powerful emergency lighting. More vehicles were arriving and parking along his road, mostly thrill seekers, Gray expected.

Strangely enough, the garage end of the house appeared virtually untouched, strange because Gray's automobile which had been parked there lay on its side crushed against a large tree.

Refusing to be hurried along, Gray shined his light inside the garage. Pristine! Objects had fallen from shelves, but his Harley sat upright on its kick stand. Things were looking up.

Gray was allowed to sit inside an ambulance. It was a good choice. His balance was poor, his hearing degraded by constant ringing, and his entire body was developing a dull ache.

Trooper Bill, last name still unknown, pushed a worried face into the ambulance. "How are you doing, Mister Grayson?"

"OK, Bill. Call Bucky for me, will you?"

"He's off duty, Mister Grayson, and some high powered people are on their way here."

"Call him, Bill. Tell him he doesn't have to be in uniform." Bill went away, and despite the clamor, Gray decided to doze a little.

The crime scene was isolated, dawn arrived, Gray slept in an old recliner he kept in the little damaged shop, and Bucky Larson stood guard.

Bucky roamed some, making sure that Gray's belongings did not become handy souvenirs. At Gray's request, he had locked an inside steel door that led to the cellar. Altering a crime scene was against procedure, but Gray's basement was untouched and was not really a part of the investigation.

Bucky had felt their house rattle from the explosion, but they were a mile away with a whole town of buildings to absorb the blast. His mother had called from her room, "Bucky, was that an explosion?"

"felt like one, but it could have been a sonic boom." Larson had gotten off duty at eleven PM and had been sleeping heavily.

Almost immediately he heard the fire department move out. "Guess it was an explosion, Mom. We'll find out in the morning."

He slept until Bill called. His friend's explanation got him moving, but he took time to get into uniform. In civvies, he would spend half his efforts trying to explain what he was doing on the scene.

The carnage stunned Officer Larson. The place was unrecognizable. Instead of a handsome home an ugly, rectangular bunker squatted menacingly on the naked land. Vegetation was gone and only a few stripped and broken tree trunks remained. Blinking lights on official vehicles turned the scene into a nightmare.

Gray seemed cool enough. He sat in an ambulance with a blanket wrapped around himself. He pointed to his feet, and Bucky saw that he wore a Nike running shoe on one foot and some kind of a leather dress shoe on the other.

Gray said, "I might be setting a new bomb victim style, Buck." Larson tried to laugh.

Gray shook his head as though to rid it of confusion.

"OK, first call your grandfather. Tell him I was bombed, I am all right, and that it is a local affair. Stress the last part, Bucky. It is important.

"Then see what you can do to protect my goods. Christ there could be guns, money, papers, who knows, spread all over hell."

Bucky made the call and a second to his mother. He was not surprised when she said she was on her way.

Confined by the bunker, not too much of Gray's personal stuff had been blown away. An oddity struck Bucky's light, and he went into the field to investigate. Sticking butt up with the muzzle driven into the dirt, as though marking a grave was a scoped rifle. Bucky pulled it free to discover what looked like part of a silencer attached to the muzzle. The stock had been shattered, but the action worked. *.44 Remington Magnum* was engraved along the barrel. Gray had not shown him this gun. Whether legal or unlicensed, the suppresser could only raise questions, so Bucky held the piece close along his leg until he found a corner of Gray's shop that would hide it.

Questioning Grayson had to wait until the man woke up. The important people did not like it, but they filled in by bothering the forensic teams and obliging or avoiding TV and paper journalists. When Gray came stretching and breathing deeply from his toolfilled lair, they had a multitude of questions.

Gray used their portable john, accepted coffee, thanked fire fighters and ambulance crews. He threatened an intrusive AM radio reporting team, and they rushed their scoop for immediate broadcasting. Then Gray sat for the official investigators.

He had little to tell them. He had been asleep, heard noise, saw lights, and got blown halfway to hell.

Who? Somebody who had owned an atom bomb it looked like.

But who? The only connection he could make was to a motorcycle thief killed trying to steal his cycle. The man was said to be part of some radical militia out west somewhere.

The F.B.I. was there hoping to make a terrorist or interstate connection which would let them in.

Who was Grayson? An obscure retired clerk type from Congressional Budget. No one would have an interest in him. Grayson had been a medium ranked courier since the 1950s. Just another face in the crowd. The militia guy had to be the connection.

But the bunker? Nobody owned such a structure. Nobody.

The answer was simple enough; Grayson harked back to the bomb shelter days. Having such a place had been his dream since the 1950s when the government had offered plans for all kinds of protective shelters. So, he had saved and had it built. Not for the purpose that had spared his life, but there ought to be a lesson here. Mister Grayson intended to rebuild, only stronger.

Anne Larson returned with Doctor Fritz in tow. The physician lugged an ancient, black home visitation bag. He looked into Gray's eyes, listened to his chest, rubber hammered a few joints and declared him apparently sound. Gray felt massively unsound, but he made no announcement.

Anne insisted he come to their home. Gray whispered to her that it might not be over yet. Brad Gilmore's daughter understood. For now, Gray was safer on the scene and surrounded by officialdom.

By nightfall the crowd was gone. Lights had been restored to the shop and garage. A plumber had a bathroom working, and a bunch of Bucky's friends had restored a lot of order within the bunker. Gray would spend the night in the shop, then move back into his house. The place was battered, perhaps beyond practical repair, but it could do for now.

The explosion dominated local news, but what was one house on a national scene? Networks passed.

Cars came down the road, idled, then drove on. A lot of photographs were taken.

Bucky said, "We ought to run all of those plates. The guys who did this will want to look."

"They could be female bombers."

"What? Women don't do this stuff, Gray."

"Guess you are right, but who does?"

Bucky was hesitant. "Maybe the people you built this bunker against, Gray."

Gray shrugged, "Et tu, Bucky? I've told everyone that's asked that this house was a cold war idea that I always wanted to do. I got the scheme back in the late 1950s, when everybody wanted a fallout shelter. There are no bad guys from my past, Buck. Trust me on this one."

"Then it almost has to be the Pittsburgh Militia."

"If it is, they are stupid people. This was a fertilizer bomb, of course, and a huge one. All the way from Pittsburgh? For the guy who owned the motorcycle? Thin, Bucky, really thin. Work through it again."

"Well, that's what you told the FBI and . . ."

Larson's eyes lit up. "Dillard and Crowder were local, and you laid them up bad.

"Local militia, sure as shooting, Gray."

"It makes more sense, but it is still thin. Who would care that much?"

"Buff Bolinger would care."

"You mentioned him before."

"He is the County Militia. Bolinger could have taken it personal." Bucky paused, "Maybe they can trace the truck. If it's from around here we might develop a case."

Gray said, "Look over here." He pointed down the bank almost to the river. "What do you make of that?"

Larson started down, but Gray held him back. "It's what's left of a light bar, Buck. The kind they put on pickups, and it isn't mine."

"It doesn't look completely blown up, just all bent up. Maybe somebody threw it there. The trees would have covered it."

"Yeah, or maybe it got ripped off when the explosive truck was driven onto my porch."

"Damn, that could be."

"You can report it, Bucky; maybe it will tell something."

Gray felt a little dizzy and sick to his stomach. Time to sit down again. He needed the time anyway. The investigators could fumble around. They did not know about the dead militia Snipers being set on him, but Bolinger did, and Bolinger had decided on strong measures. Only the bunker, perhaps unsuspected or at least underestimated by Bolinger, had saved Grayson.

Gray needed a day or two to steady down and maybe another to reconnoiter. Then he figured he would just clean this mess up once and for all.

17

Bolinger stared at the newspaper photograph. The bunker looked like a World War II fortress. He could never have imagined such a mass of iron and concrete hidden within an ordinary house.

Who and why turmoiled Bolinger's mind. What kind of a man built and lived within such a place?

A survivalist kind of loner left over from the cold war bombing fears, the newspapers claimed. Fobbs had muttered about some kind of bomb shelter or something, but Bolinger had imagined an underground place that you ran into when warnings were broadcast. This thing--this was Maginot line stuff. And worst of all, Grayson was not dead. From all the reports, Grayson was not even hurt.

The immensity of the bomb was luridly discussed. An amateur photo of Grayson's house was shown beside what was left. Some of the builders had appeared, and their comments were rich reading. Three foot thick walls, armor plate shutters, good God. The only good news was that because no one was even injured by the huge blast no network news featured the story.

Charlie Fobbs was scared into serious shakes. He had to take time off from work, and tried to hang around the gun shop. Fobbs was coming apart, but Buff had no toleration for the man's weakness. Bolinger's veto was harsh.

"Get the hell to work, and act natural, for Christ's sake. Do what you always do, eat what you always eat, talk about the bombing if it comes up, but don't know anything beyond what the papers say."

The bombing was the gun shop's major topic for two days. Every militiaman came by or called in. The *Patriot-News* hinted of a revenge motive by a Pittsburgh militia because of the death of one of their members while stealing Robert Grayson's motorcycle.

A reporter who interviewed Colonel Buff Bolinger, Commanding Officer of the County Militia, speculated that whoever had destroyed Grayson's home had also blown up the two local militiamen, but Bolinger dismissed the connection. Although liking the rumor, Bolinger did not want it coming from him. He pointed out that according to the reporter's own newspaper, the police described entirely different explosive types. The two men were blown up by military explosives, the Grayson place by a fertilizer bomb.

The truck was currently a mystery. Bolinger had chiseled away the I.D. plate, but eventually Bags Dillard would leave the hospital and discover that his pickup was gone. When that happened, Colonel Bolinger would snarl that his people were not stupid enough, or rich enough to use their own trucks, and it was clear that someone was trying to place guilt on innocent parties.

What would Grayson do? Buff Bolinger supposed that the man already knew who had twice tried to kill him. The dead Snipers could have spilled their guts before dying, but even that was not necessary. Militia members had attempted to assault him, County Militia Snipers had tried for him, therefore, the same enemies had blown up his house. Who could reason otherwise?

Grayson would come for the militia leaders; that was what he would do, and Buff Bolinger had no idea what to do about it.

Buff mailed off an anonymous letter warning Grayson to back off, but he had few hopes for it.

More Sniping was impractical. Hell, he did not have any Snipers, and he had no idea where Grayson had moved to.

He did not even know what Grayson looked like. The paper had a long distance shot of Grayson, but it could have been anybody. The inquiring reporter told that Grayson would not talk to them directly. His interview was by phone. Their photographer came back with trash. The man claimed it was like trying to photograph Wilson, a popular TV character who talked from behind a fence. Grayson always had a hat pulled down or a hand rubbing his nose. He wore sun glasses all of the time. Who cared, the reporter decided, the story was dying anyway.

Four days after the explosion, Bolinger discovered what Grayson looked like.

The gun shop was busy with a customer and the usual hangers about. Bolinger was closing a deal on a used Mossberg 500 model shotgun to an older man who said he planned to shoot the son of a bitch who had come on his land unannounced.

Buff heard a lot of that kind of talk, but nobody ever shot anyone. Bolinger took the filled-out purchase form and prepared to do his part.

For a moment, the name Robert Grayson meant nothing. When it registered, Bolinger froze and his pen laid a long ink streak across the form. His eyes jumped to the buyer's face and encountered the coldest eyes he had ever dreamed of seeing. A corner of the hard mouth twitched as though suppressing a smile, and Bolinger knew that his agitation was understood by Grayson.

His mind racing, Bolinger got his pen going again. He wondered if he could shoot Grayson--his pistol was in its buttoned down holster at his waist. Everybody would see him kill the unarmed Grayson, but when would he have another chance? It then occurred to him that Grayson had planned the confrontation, and as sure as hell the man was armed. Bolinger cursed the military holster that took all day to draw from.

Grayson paid and took his shotgun. He operated the pump action and asked, "Where's the best place to shoot a man when you are close together?" While Bolinger struggled for words, Grayson added, "I heard you were a militia colonel, so you'd know about things like that."

Bolinger made his voice strong. "Always hold for the center of mass."

Grayson's eyes fell to Bolinger's prominent gut. "I'll keep that in mind." Then he left.

Bolinger realized his face was sweat soaked, Christ, even with the air conditioning on. He stepped into his toilet and dried himself. In his mirror he appeared old and drawn.

Grayson was mocking him, making him half sick with fear and worry. When he was ready, Grayson would step out, probably with that damned Mossberg, and let him have it, in the gut, right where he had looked. Just like Babe Ruth had pointed to the right field bleachers before knocking a home run over them. Oh God!

And Bolinger still did not know what to do about it.

Gray had to know what Bolinger looked like. If finding out could also help make sure that he had a guilty man, so much the better. He had watched Bolinger closely. The shop owner had shown no signs of recognition until he read the name. Then he had nearly collapsed.

A normal reaction would have been to say, "Hey, you're the guy whose house got blown up." But the colonel could not handle it.

Gray had decided to grind Bolinger's guts a little more, and his question about where to shoot a man should have done it. He would let Bolinger eat his liver for a while--but only a short while.

The letter was addressed only to Robert Grayson, city, and state. It had been dropped into local mail, Gray could tell by the postmark, and sent by someone who did not know his post office box number. The letter was typed and said:

Grayson
Do not hunt us or
we will kill your woman.

There was no signature.

His woman? He had seen Anne Larson only a few times. Once at his home, once at hers, their trip to Mayo, and some meals since. But, there was no other woman; the letter writer either meant Anne, or he was simply shooting blind, hoping for reaction.

Gray thought about it. The bad guys could expect that he would have a friend, and that would make their threat intimidating. If not, they had lost nothing. Most men might have been concerned. Grayson just decided to move his plan along more rapidly.

Gray had found the letter waiting when he had returned from buying the shotgun. The trip to the gun shop satisfied his need to believe that Bolinger was behind the Sniping and the bombing. The colonel's poorly disguised fear had been a convincer.

His next stop was in the city. He had worked through the public library's microfilm copies of the *Patriot-News* tracing the County Militia to their roots. Colonel Buff Bolinger's efforts were documented with numerous pictures of Bolinger. Invariably, his aide, Major Charlie Fobbs stood at the colonel's elbow. Only those two received more than passing notice. One militia photo did include two men in Sniper garb. Gray remembered their faces.

As Bucky Larson had suggested, the militia was not much. As Brad Gilmore reminded, cut off the head and the body would die.

Bolinger would go, and Fobbs, too, if he was handy. Fobbs probably knew things, but without Bolinger to haul him along, Gray expected the adjutant would resign his rank and disappear. The County Militia, Gray decided, had enjoyed its last meeting.

During the days, Gray met with insurance adjusters, police investigators who still poked at rubble, and his original builders with whom he wrangled over specifics of rebuild.

Gray intended to increase the thickness of his outer walls by an additional foot of concrete. His call to his steel plate provider disclosed the availability of one inch thick armor that was much tougher. The vastly heavier steel would require motors. They discussed servos that could close one shutter or all with the flip of a switch. "T" sections at shutter tops and bottoms with interlocking at shutter closures was agreed upon. The overlaps would prevent the buckling that had destroyed his old shutters. Gray was satisfied that if ever needed, his new security shutters would hold.

Anne came with lunch, and they dined on top of the bunker, dangling their legs and looking across the river. Grayson examined a qualm or two about perching exposed as a sparrow for any interested marksman but decided to hell with it.

His cleaning crew had already cut away the ruined trees and planted pine seedlings. Anne spoke about it.

"It will take years for those trees to grow back, Gray."

"Too long. I'm having some five year old plantings brought in. That should cut the time down."

Gray kept his voice level. "Do you like it out here, Anne?"

"This is a wonderful spot, Gray. You could have a small dock and a boat if you wanted to."

"I have intended dropping a dozen loads of sand and making a beach. Years ago a buddy and I spent days moving all the loose rocks from under a big tree limb. That tree just became mulch, unfortunately, but we used to have a rope swing there, and we could swing out and drop into the clear water. That's where I'll put the beach."

"You swim in the river?"

"Always did and still do, sometimes."

"It isn't too clean, Gray."

Gray laughed, "Cleaner than it used to be, but our river will never be perfect. I accept what it is and dive in."

"You could have horses in your field."

"Do you ride, Anne?"

"I used to, but renting horses is expensive."

"I could like having horses, but someone else will have to take care of them."

"That is easy, Gray. Teenage girls love horses. If you paid them a little and let them ride, you could have a dozen to choose from."

"I'd like the horses, but I don't know about the teenagers."

Because it was Friday, the workers left early. When Anne and all were gone, Gray went to his cellar.

With electricity out, he had to use a pair of battery lanterns. He locked his door at the stairhead and went to his vault-doored arms room. Inside, he closed and locked his inner door, then opened his secret room.

Below ground, with the foot thick concrete above, the munitions had barely jiggled. Gray began to select and lay aside what he would use.

It could be easy to Snipe Bolinger, but explosives seemed in vogue, and as had been demonstrated, blowing something all to hell made a powerful impression. Gray wanted the surviving militia to be impressed and convinced that they might better enjoy other activities.

When he had described the carnage the Bolinger bomb had caused, Brad Gilmore had been impressed.

"A man who would go to those lengths is dangerous, Gray. My God, the mixing alone must have been a monster of a task. You can't be sure he did it alone."

"My estimate is that he had one helper, his right-hand man, a pipsqueak named Charlie Fobbs."

Gilmore sighed, "There is only one solution, Gray. You know that, of course. If you do not make sure, you will never be wholly certain that they won't come again. As long as they live you will wonder."

"That's the way I see it."

"Well, you know how if anyone does. Keep me posted, Gray."

Gray worked within his secret room. If there was an explosion while he worked, it would all go up, but what would he care? His atoms would be well distributed, anyway.

Gray cut lengths of detonating cord and squeezed foot long lengths of plastic explosive around each length. He wrapped each explosive length in clear plastic and taped the protection in place. He tested with a short length of two inch diameter PVC pipe. His explosive chains had to easily slide through.

When finished, he had three chains. Gray folded each explosive chain into a foot long, easy-to-carry bundle.

Gray planned to explode his charges from a distance. His device resembled a garage door opener with a cover and a safety pin to prevent an accidental squeeze.

He placed his chains, more prima cord and extra explosives within a pair of laundry bags and slid the detonating device into a pocket. His clothing was dark, and it would soon be night. His nondescript rental car was waiting. Gray loaded his material and was ready.

Criminals wanted guns, so gun stores were subject to burglaries. Gun shops invariably had small, heavily barred windows and steel doors. Their roofs and walls were not easily penetrated, and all kinds of alarms were employed. Some shops placed their more expensive weapons in a vault each night and returned them to their show cases each day.

Gray knew about such elementary defenses, and he could have gone through them. He did not need to. If properly used, he had enough explosive to junk Bolinger's establishment and whoever was inside. In Grayson's plan, the building's inherent strength would work to his advantage.

He came out of Bolinger's woods, no more than a swiftly drifting shadow. Gray gained the gun shop's roof by using a woodpile as a ladder.

The laundry bags were heavy and awkward, but Grayson knew how to shift such loads. He gently placed each on the shop's flat roof.

Gray avoided sudden jars. Bolinger could be clever enough to have sensors. These days, everyone interested could get motion, sound, or pressure sensing devices that could be rigged to transmit alarms. Fortunately for Grayson's type of work, such alarms had to be coarse enough not to respond to jet aircraft pressure or passing shadows of blown vegetation or flying birds. Gray still moved lightly and avoided making loud noises.

Across the roof, Gray placed block TNT. He connected the explosives with the prima cord That was itself an almost instantaneous explosive.

Of course, Gray had studied the building. The observation had been accomplished in passing. His professional eye required no more.

Bolinger's roof had three protruding vent pipes. The vents bled off sewer gases and were common and necessary on all housing. One vent handled a kitchen, another would go to a bathroom. The third, Gray could not determine. It did not matter, he would use it.

Gray lowered his explosive chains down each vent until they completely disappeared and struck bottom. He connected each chain's explosive cord to an extremely short fuse length. More explosive prima cord joined the roof explosives, and all ran to the radio signal detonator.

Gray's device was foreign, and it was old. He raised the short antenna, almost embarrassed by the design's antiquity, but despite its age, the old model detonator always worked.

Gray left the way he had come. The night was young, and he would sleep until dawn. Before the shop opened, he would be in place among the trees where he could watch for Buff Bolinger's arrival.

18

For the third night Charlie Fobbs had not slept. He was going to die. He knew it, and the realization was ripping him apart.

Buff Bolinger's words did not help. Bolinger was also scared foolish and had no idea what to do either. They had screwed up royally, that was obvious. They should have backed off when the Sniper team got blown up. Good God, they should never have started in the first place. What was Grayson to them? Nothing, that was what.

Nothing until they stirred him up. Then it turns out the guy is some kind of freak who lives inside a fort. Fobbs knew the weirdo could and would kill. The Snipers proved that, and Grayson had to have been damned good to have killed Wolfe and Tilson.

Now, just yesterday, Grayson had walked right into the gun shop and practically told Buff to get ready to die. Oh God, and he would be next. As sure as he knew about all of it, he Charlie Fobbs, was going to be killed.

His fingers could barely manage the old style military pistol belt, and the weight of the M1911A1 Colt .45 pistol at his hip provided no sense of security. He could not hit anything with it under good conditions. Today? Fobbs would have puked again if his stomach had anything left to expel.

Fobbs slid a loaded magazine into his AR15 and chambered a round as Bolinger had ordered. He put on the safety and blankly studied the rifle. What in hell were they doing? Lugging loaded rifles around in cars was against the law, and what good would it do against an enemy who had already killed their best and survived their meanest effort?

Fobbs stayed in his kitchen and well away from the door. The killer could be out there now, just waiting until he showed himself. The sudden appearance of Bolinger's truck almost at his door unhinged Fobbs, and he heard himself sob.

God, he was a mess. He clutched his rifle and forced himself out the door. His flesh cringed, but no shot came, and he made it to Bolinger's truck.

Bolinger shot him a look as they pulled away. The colonel regalia did not inspire Fobbs as it once had, and his own major's insignia felt merely foolish and play acting.

Bolinger said, "Christ, Charlie, get hold of yourself. You look like hell."

Without mortification, Fobbs heard himself sob and was aware of a tear on his cheek.

"Oh God, Buff, what are we going to do?"

Bolinger wished he had a strong answer. Grayson's appearance in the gun shop had scared him. It wiped out any lingering doubt that Grayson knew and that he meant business, but, what they should do, Bolinger could not decide.

Kill Grayson somehow, Bolinger was certain of that, but how? Buff Bolinger had never even shot at anyone, much less killed. Imagining and planning had been heady stuff. At gun sight level the fun had disappeared, and only gut grinding anxiety and debilitating indecision remained.

Visions of Grayson appearing in his shop with the Mossberg leveled trembled Bolinger's senses. What could they do?

Bolinger had a machine gun hidden away. It was an old World War II Sten gun. A stamped metal qualityless antique, the Sten could still empty a long magazine of 9mm bullets in an eye blink. Maybe he and Charlie could place themselves along Grayson's road. When the man drove abreast of them he could empty the Sten into Grayson. Fobbs could charge out and pour double ought buck or .223 rounds in to make sure. Then they could escape through the woods. Bolinger was heartened by the scheme, but looking at the mess Fobbs had become was again discouraging.

No matter, they would have to act tonight because Grayson might not wait. He would close the gun shop exactly on time, and he and Charlie would go and get the job done.

During the day, he would watch the door and have Fobbs ready with the shotgun. Grayson would not catch them unaware. Fobbs would straighten out once he understood the plan.

Bolinger felt a lot better. He was opening the shop early. A customer was coming to pick up and zero a new rifle on the shop's outdoor range. Buff wanted to be early to get targets out, the vault opened, and Fobbs squared away. God, the man was a mess.

Gray relaxed, his back against a tree with the gun shop lot before him. When Bolinger's pickup wheeled in a half hour early Gray was pleased. If innocent people had been waiting at the shop door, his detonation would have to wait. That was an explosive's greatest weakness, it did not discriminate. A blast took everyone and everything. If you did not care, plastique was wonderful, but if you did wish to include only the guilty, care had to be taken.

Bolinger and a small man Gray recognized as Fobbs left the pickup and marched directly to the shop door. Both men were dressed military in fatigues with boots bloused. Each wore a pistol and web belt with ammo pouch. Each carried an AR15 rifle.

Bolinger leaned his rifle against a wall and went to work on the shop's multiple locks while Fobbs fingered his AR15 and stared at the surrounding forest. He had no chance of seeing the dark garbed and motionless Robert Grayson.

Fobbs stepped on Bolinger's heel hurrying through the door. Bolinger said, "Damn it, Charlie, settle down," and hurried to shut down the burglar alarms, one of which sent a silent alarm directly to the state police sub station.

"He could be right out there in the trees, Buff."

"Sure, Charlie, and what would he do, start a firefight? The man's not a fool, Charlie. He'll wait until he thinks it is just the right time." Bolinger's words made them both feel better. Grayson would have to be stupid to charge the gun shop with his shotgun.

Bolinger chose paper targets and handed them to Fobbs. "Go hang these down range, Charlie. They'll be here and shooting directly."

Fobbs hesitated, "He could be out there, Buff."

"Just go do it, Charlie." Bolinger let his impatience show.

Fobbs went to the back door, laid aside a heavy iron bar and released the dead bolt. He peered out, but the woods revealed nothing. Sucking in air, his knees trembling, Fobbs headed for the range butts. Only a hundred yards or so. He would be back inside in minutes, and surely Grayson would not shoot him when Buff was safe inside. Would he? Fobbs broke into a shambling run, his rifle and the targets heavy in his hands.

Gray rose and stepped behind the protection of his tree. He did not anticipate a lot of stuff flying, but some would.

He gave Bolinger time to turn off any alarms. Maybe Bolinger or Fobbs would go to the rest room or be at the sink making coffee. That would be right in the explosion.

Gray removed the safety from his detonator and opened the cover. A last look around showed nothing moving. He held the device around the tree trunk and pressed the trigger pad.

Inside the shop, Buff Bolinger had just opened the door of his gun and gun powder vault. The room was heavily armored against theft and was kept dry by an air circulator that sucked air into the vault through small wall holes into the shop and exhausted out a common plumbing vent through the roof. Bolinger sniffed the vault air. It smelled a little dank and stale, as if the roof vent was blocked or something.

Grayson's roof receiver detonated a blasting cap that instantaneously exploded the TNT deposited on the shop roof. Delayed by the short fuse, the plastique chains detonated an instant later.

The initial blast began the roof's collapse. Bolinger's mind might have registered a pressure increase, but the three explosive chains blew away any possibility of realization.

The chain blasts blew horizontally. The explosions decimated the shop's interior. They blew out windows and both doors and ruptured the shop walls. The most effective was the chain in the storage vault's vent. The explosion wiped away all that had been Buff Bolinger. It also ignited the gunpowder stored in the vault.

The chain explosions halted the roof's downward collapse, reversed its direction, and blew its pieces into a fountain of bits and shattered plywood.

As the thunder of the multiple blasts reverberated across the ridges, Gray took his look. The ammunition explosion still crackled as overheated cartridges fired, but loose ammunition burning held no danger. Without compression, the bullets moved only a few feet.

The gun powder blast was pure bonus, and Gray expected that for a few hours the destruction would be considered accidental. Not for long, however, an experienced investigator would recognize multiple sources, and the fire marshal would not be fooled at all. Before then, Gray planned to be far away.

With the building down, Gray could see further. Something glinted and something moved beyond the shop ruins. Gray shifted position a little to see more clearly.

Holy hell, there was Charlie Fobbs crawling like a snake toward a shooting range backstop. Unbelievable, except that unpredictable things often happened.

Gray supposed that Fobbs had gone straight through the shop and out the back door. Gray drew his .380 and trotted down the slope and around the now blazing ruins to finish the job.

Charlie Fobbs had been blown flat. His rifle was gone and his uniform was nearly torn from his back. Knowing only fear, he crawled blindly until he saw the shooting butts almost within reach and blocking his way. He whirled to face the gutted and burning ruins. He too would be blown to bits, just like Wolfe and Tilson and now Buff. Grayson blew everybody up. Fobbs' bowel released, and he felt the hot wetness in his crotch.

Fobbs fumbled out his pistol and managed to chamber a round. His eyes searched frantically, and there he was! A black clad figure running straight at him, already past the gun shop and coming fast.

Fobbs pointed his pistol, but the figure had melted from sight. Grayson was coming for him, to blow him to bits like the others.

Well, he wouldn't get Charlie Fobbs. Fobbs felt himself turn clever. Grayson could be close, and he had no time to waste.

Charlie Fobbs forced the pistol's muzzle into his mouth and pulled the trigger.

Gray saw the back of Fobb's head explode before the muffled report reached him. He rolled to his feet, pulled up short, then came on at a walk.

Fobbs had been thorough. Sometimes a bullet strayed or stayed inside a skull. Fobbs' had gone through, and concussion had also blown the man's eyes clean out of their sockets. Gray had seen it before.

He got out his detonator and stuffed it into Fobbs' pocket. His voice was cynical. He said, "Case closed."

Vehicle tires screeched, and Gray hit the ground as Bucky Larson's cruiser swerved into view and skidded to a halt.

Larson punched his radio as he stopped. The flaming ruins held his eyes, but possible movement at the periphery made him look. Someone was lying out there, but there was no movement.

He called in describing the disaster. Then he sat for an extended moment wondering what in hell was happening to his county.

19

James N. Waverly came on Monday. He had a government car and a burly driver with an ill-concealed pistol in his armpit. What a dullard to handle covert work.

Gray had agreed to a meet, far out of town, in a gas station and Handy Mart parking lot.

As usual, Waverly's head was glowing with angry heat. He said, "God damn it, Grayson, what in hell are you doing up here?"

Grayson's calm had always infuriated Waverly, and it did now.

"I am just a victim here, Waverly. Read your reports. The bomber was a guy I never met. Some kind of crazy militiaman who blew his own head off. Hell, he had the detonator on him, and a cop was on the scene while boards were still falling. What do you want from me?"

"People are asking into your background. They want to know how a nothing of a government courier can afford a place like you had, and . . ."

Gray interrupted. "You take care of those inquiries, Waverly. That is what you are hired for." Grayson's smile was cold. "You would not want me exposed, now would you, Waverly?"

Waverly's temperature dropped degrees. He said bitterly, "That is still blackmail, Grayson."

"And still effective. Go back to Washington. Do your job. I am retired, and I expect to be left alone."

Waverly went away. He wondered for the thousandth time if Grayson still had the incriminating tapes. He had to assume he did. Damn Robert Grayson!

Gray said, "A good thing about doing it all over again is that the workmen know what I want."

He and Anne had watched the pouring of the new and thicker outer house walls.

She said, "I hope no one blows it up again, the insurance company is still weeping."

Bucky heard their laughter and came over, carefully lifting his polished boots over the rough surfaces.

Gray said, "Your mother and I are going down to Saint Thomas in the Virgins for a couple of weeks. You are welcome to come down."

Bucky checked Grayson's arm lying familiarly around his mother's waist.

"I guess not, this time, anyway. Rooms cost a lot down there."

Gray said, "I've got a small place, not fancy or anything, or you could stay on the boat."

Bucky goggled. "You own place on Saint Thomas, and you're renting a boat?"

"No, I own the boat. It's usually in charter. It's a 3400 Gemini, a little roomier than your father's."

"You own a boat?"

Gray laughed, "Yeah, I own a boat."

Bucky said in mock seriousness, "Mom, marry this guy. He has his own boat."

Anne's laughter was contagious, and they were all grinning. Her eyes on Gray were warm. "We'll see, Bucky, we'll see."

Later, Gray said, "There will always be secrets between us, Anne, deep and deadly things that can never be told. You know that, don't you?"

She sighed, "I know, Gray. I lived with my father's secrets, remember?

"But Gray, those secrets are not between you and me, just as they were not between mom and dad. I know how to handle those things."

She stuck out her firm and, to Gray's eyes, interesting bosom in pretended pride. "And you will not find many women who can truly make that statement."

Gray laughed in solid satisfaction. "Anne, I admit that you are special, and I have only one comment to add."

Her glance was questioning, "And that is?"

"I'm glad Bucky is not coming along on this trip."

Their laughter turned the heads of the sweaty concrete workers who cheered and raised their iced beers in salute.